# VOIP TECHNOLOGIES

# VOIP TECHNOLOGIES

*Editor*

**Jeremy Weissberg**

scitus academics

**VoIP Technologies**
Edited by **Jeremy Weissberg**

ISBN: 978-1-68117-608-6
Library of Congress Control Number: 2016946463


SCITUS Academics LLC,
616, Corporate Way, Suite 2, 4766,
Valley Cottage, NY 10989

www.scitusacademics.com



**Notice**



Printed in United States of America on Acid Free Paper

# PREFACE

Voice over IP (VoIP) is a methodology and group of technologies for the delivery of voice communications and multimedia sessions over Internet Protocol (IP) networks, such as the Internet. Other terms commonly associated with VoIP are IP telephony, Internet telephony, broadband telephony, and broadband phone service. VoIP is undoubtedly a powerful and innovative communication tool for everyone, voice communication over the Internet is inherently less reliable than the public switched telephone network, because the Internet functions as a best-effort network without Quality of Service guarantee and voice data cannot be retransmitted. Because of the bandwidth efficiency and low costs that VoIP technology can provide, businesses are migrating from traditional copper-wire telephone systems to VoIP systems to reduce their monthly phone costs.

*VoIP Technologies* introduces research strategies that address various issues with the aim of enhancing VoIP quality. VoIP solutions aimed at businesses have evolved into unified communications services that treat all communications—phone calls, faxes, voice mail, e-mail, Web conferences, and more—as discrete units that can all be delivered via any means and to any handset, including cell-phones. VoIP allows both voice and data communications to be run over a single network, which can significantly reduce infrastructure cost. We hope that you will enjoy reading these diverse studies. This book will provide a lot of useful information about current VoIP technology research aimed at practitioners and researchers as well as advanced graduate students.

# TABLE OF CONTENTS


Chapter 1 Vertical Handover Management for VoIP Session over Brodband Wireless Networks 1

Chapter 2 A study of artificial speech quality assessors of VoIP calls subject to limited bursty packet losses 31

Chapter 3 A Topology-Aware Relay Lookup Scheme for P2P VoIP System 71

Chapter 4 An Efficient Scheduling Scheme to Enhance the Capacity of VoIP Services in Evolved UTRA Uplink 91

Chapter 5 Re-Engineering Campus-Wide Internet Telephony Using Voice over Internet Protocol 115

Chapter 6 QoS-aware composite scheduling using fuzzy proactive and reactive controllers 133

Chapter 7 Testing and Analysis of VoIPv6 (Voice over Internet Protocol V6) Performance Using FreeBSD 179

**Chapter 8** **Efficient Control Channel Resource Allocation for VoIP in OFDMA-Based Packet Radio Networks 195**

**Chapter 9** **Solutions Comparison towards Voice Services Implementation for Operators Starting LTE Deployment 229**

**Chapter 10** **Improving Management Performance of P2PSIP for Mobile Sensing in Wireless Overlays 249**

**Index 285**

1

# VERTICAL HANDOVER MANAGEMENT FOR VOIP SESSION OVER BRODBAND WIRELESS NETWORKS

**Muhammad Niswar[1], Shigeru Kashihara[2], Suguru Yamaguchi[2]**

[1] Department of Electrical Engineering, Faculty of Engineering, Hasanuddin University, Makassar, Indonesia

[2] Graduate School of Information Science, Nara Institute of Science and Technology, Nara, Japan

## ABSTRACT

Today, wireless LAN (IEEE802.11g/n) has been the dominant wireless network that can provide high data rates in a limited coverage area. While emerging mobile WiMAX (IEEE802.16e) can provide a wireless broadband access to mobile users in a wide coverage area. These two different technolgies will co-exist while complementing each other in some regions, hence, a mobile node (MN) with dual interfaces traverses and executes many handovers (HOs) between 802.11g and 802.16e. Meanwhile, there is a huge demand for Voice over IP (VoIP) service over wireless networks. However, VoIP is a delay and loss sensitive application, hence, VoIP session is more likely to be deteriorated during HO between 802.11g and 802.16e. In order to maintain quality of VoIP session during HO, we proposed Vertical Handover Management (VHM) for VoIP session that focuses on HO initiation and decision strategy according to the wireless link condition and congestion state of wireless networks. The VHM exploits request to send (RTS) retries and round trip time (RTT) between an MN and an access point in an 802.11g interface as well as a carrier to interference noise ration (CINR) level and MN's queue length of an 802.16e interface as HO triggers. We then conducted simulation experiments to evaluate the effectiveness of our proposed VHM using QualNet 4.5. Our simulation results show that our proposed VHM can preserve the quality of VoIP session during such HOs.

## KEYWORDS

Handover Management; Handover Triggers; Single/Bi-Casting; 802.11; 802.16e; VoIP

## 1. INTRODUCTION

Wireless networks will consist of various wireless network technologies and mobile users will traverse several coverage areas of different wireless access technologies. Recently, mobile WiMAX (IEEE 802.16e) has gained serious attention as a means of providing wireless broadband access to mobile users in a wide area, and it provides QoS for various applications. On the other hand, IEEE 802.11g has already been widely used to provide high data rates in a limited area. Therefore, in the near future, since 802.11g and 802.16e networks are intermingled while complementing each other, a mobile node (MN) will be likely to execute many handovers (HOs) between 802.11g and 802.16e networks with different IP subnets (e.g. two networks that administrated by two different ISPs).

Meanwhile, there is a huge demand for Voice over IP (VoIP) service over wireless networks. However, delivering VoIP over wireless has many challenges because VoIP is a delay and packet loss sensitive application. In a mobile and wireless environment, typically, two main factors degrade VoIP quality over the wireless network: 1) degradation of wireless link quality; and 2) wireless network congestion. First, because an MN moves freely, the communication quality degrades due to the fluctuation of the wireless link condition. Second, as VoIP is a bi-directional communication, an access point (AP) of 802.11g and a base station (BS) of 802.16e become a bottleneck with the increase of VoIP calls. When an MN traverses several APs and BSs, an MN needs to detect the degradation of VoIP quality and HO to the best available wireless network.

In order to detect the degradation of wireless link quality as well as a congestion of wireless network in a heterogeneous wireless environment, we cannot only rely on one HO trigger such as Carrier-to-Interference plus Noise Ratio (CINR) level. This is because in vertical HO, we cannot directly compare the CINR level of different wireless technologies since each technology has different characteristics in terms of MAC and physical layer. For instance, they might have the same CINR level but differ in data rate. Therefore, it is a challenge to develop a vertical HO management (VHM) considering suitable HO triggers in order to achieve seamless HO between two different types of wireless networks. In this paper, to preserve VoIP quality during HO, we then propose VHM for VoIP over the intermingled 802.11g and 802.16e networks considering the wireless link condition and congestion state of networks.

## 2. RELATED WORK

So far, several studies on the vertical HO schemes have been conducted. References in [1, 2] proposed an algorithm for vertical HO between 3G and WLAN. Authors in [1] use the Received Signal Strength (RSS) with estimated lifetime to execute the HO. On the other hand, authors in [2] use a dynamic RSS threshold that calculated and compared with the current RSS to execute the HO. Reference [3] proposed a link trigger generation method based on

RSS. The author proposed a smart triggering scheme by determining link status transition including Link-Up (LU), Link-Coming-Up (LCU), LinkGoing-Down (LGD) and Link-Down (LD). Each link state have RSS threshold for transition. These link status will trigger the SIP session mobility protocol to execute the HO. Reference [4] proposed a cross-layer based polynomial regression predictive RSS approach with marcov decision process for optimal selection for HO in heterogeneous network. The proposed scheme consists of two primary phases, i.e., the phase of predictive RSS with a hysteresis and the phase of determining optimal target network in which to HO using marcov decision process. The author evaluated the proposed method using numerical analysis and compared various performance metrics including number of vertical HO and network throughput. Author described the proposed algorithm through mathematical analysis and evaluated through simulation. The proposed method can avoid the ping-pong effect using predictive RSS with hysteresis. Reference [1-4] schemes exploit RSS as an HO trigger, which cannot represent the received signal quality because it is only indicate the strength of received signal excludes noise, interference, and other channel effects.

Reference [5] proposed an algorithm for vertical HO between 802.11 and 802.16 hybrid networks exploiting data rate and channel occupancy as HO triggers and implemented them on top of the MAC layer. The author uses downlink Signal to Interference Noise Ratio (SINR) for wireless connectivity trigger as well as data rate and network load for performance trigger. The author validated the proposed scheme in two scenarios, i.e., singleMN moves from WiFi-WiMAX-WiFi and system capacity evaluation. The simulation scenario limited to longlived FTP application for movement scenario. There is no evaluation for real-time application such as VoIP.

Authors in [6] introduced a mobile relay node as an intermediate node between 802.11 devices and 802.16. base station (BS) to convert 802.16 signal received from BS to 802.11 devices. This approach is inefficient because it requires additional intermediate node among 802.11 access point (AP), 802.16e BS and MN.

The most recognized study is Media Independent Handover (MIH) that was proposed by the IEEE 802.21 working group [7]. MIH is intended to facilitate HO and interoperability of various other network technologies. However, deployment of MIH requires additional layers to the existing protocol stack of mobile device and infrastructure networks leading to cost inefficiency. Therefore, in this paper, we focus on an end-to-end VHM that only requires some modification on end-node, and then our proposed method aims to preserve VoIP quality during HO between networks that managed by different organization.

## 3. HANDOVER INITIATION

HO initiation is a process to triggering HO according to some parameters/specific conditions. Since we intend to develop a multi-homed MN-initiated HO scheme where MN has two different wireless interfaces (IFs), i.e., 802.11g and 802.16e networks, we need to determine the HO triggers for each different IF that can be obtained at the MN side. Our proposed VHM aims to preserve VoIP quality during HO; hence, we need to consider several HO triggers indicating both wireless link condition and congestion state in both 802.11g and 802.16e networks. The following subsections describe our proposed HO triggers and the motivation behind selecting them.

### 3.1. Handover Triggers for 802.11

As VoIP is a bi-directional communication, AP becomes a bottleneck with the increase of VoIP calls. That is, VoIP packets to MNs are liable to experience large queuing delay or packet loss due to increase in queue length or buffer overflow in the AP buffer because each MN and AP has almost the same priority level of frame transmission by following the CSMA/CA scheme. In our previous study [8], we found that Request-to-Send (RTS) frame retries and Round-Trip Time

(RTT) between MN and AP can be potential MN-initiated HO triggers for 802.11g. We briefly describe these triggers in the following subsections.

### 3.1.1. Frame Retries

In the IEEE802.11 standard, a sender confirms a successful transmission by receiving an ACK frame in response to the transmitted data frame. When a data or ACK frame is lost, the sender periodically retransmits the same data frame until achieving a successful transmission or reaching a predetermined retry limit. When frame retries reach the retry limit, the sender treats the data frame as a lost packet.

Our previous study [9] shows that frame retries can promptly and reliable detect the wireless link degradation with compared to RSS. However, current WLANs employ a multi-rate function that can change the transmission rate according to wireless link condition. If the transmission rate is dropped by the multi-rate function, a more robust modulation type is selected and thus data frame retries are further decreased. As a result, an MN cannot properly detect the degradation of wireless link quality only from data frame retries in multi-rate WLANs. We then consider RTS frame retries as an alternative trigger for data frame retries. As an RTS frame is always transmitted at the lowest rate (e.g., 6 Mb/s in 802.11a/g and 1 Mb/s in 802.11b), an MN can appropriately detect the change of wireless link quality. In our previous study [8], we found that an RTS retry ratio of less than 0.6 satisfies adequate VoIP call quality, and we suggest it as a threshold value for initiating HO in this study.

### 3.1.2. AP Queue Length

With the increase of VoIP calls in a WLAN, the AP queue length increases because all traffic to and from the WLAN has to go through the AP. The 802.11 CSMA/CA mechanism provides the AP and MNs with the same number of transmission opportunities. Suppose we have number of MNs (Nmn) transmiting upstream traffic, the MNs have an Nmn/(Nmn + AP) share of bandwidth (BW) while the AP has only a 1/(Nmn + AP) share of BW for transmitting downstream traffic. Moreover, the AP typically needs to send more packets than the MN; hence, each packet routed to the MN (downlink transmission) and queued in the AP buffer may experience a large queuing delay or packet loss due to asymmetry of BW share. Consequently, the queuing delay and the packet loss due to buffer overflow severely affect the VoIP quality of MNs. However, the IEEE802.11 (a/b/g/n) standard unfortunately does not provide a mechanism that can inform MNs of the AP queue length. Therefore, to preserve VoIP quality, an MN needs to detect the congestion of the AP by itself. In our previous study [8], we found the significance of the AP queue length and proposed a method to estimate AP queue length based on RTT between MN and AP (W-RTT). We found that W-RTT < 200 ms satisfies adequate VoIP call quality and we suggest it as a threshold value for initiating HO.

## 3.2. HO Triggers for 802.16

802.16e supports high data rates and multi service types; hence, it is a strong contender for wireless broadband access technology to support real-time applications such as VoIP over wireless networks. Although 802.16e supports multiple service types, often only Best Effort (BE) service type is made available at the initial phase of deployment. The BE scheme has no classification and priority to delay and/or loss sensitive applications; hence, the VoIP application must contend with various types of applications for obtaining transmission opportunity in 802.16e. Since we deal with a VoIP application which is bi-directional communication, we need to consider some metrics indicating both downlink and uplink transmission condition. We need to determine the HO triggers that can be obtained at the MN side.

In this study, we propose a combined use of two metrics, i.e., downlink CINR for indicating downlink transmission as well as the wireless link quality and MN's queue length for indicating uplink transmission and as well as the congestion state of the 802.16e network. We describe the motivation behind selecting these metrics as an HO trigger for MN-initiated HO of the 802.16e. We describe these metrics in the following subsections.

### 3.2.1. CINR Level

Many HO technologies employ RSS as an HO trigger, which can indicate the wireless link condition. RSS provides a simple indication of how strong the signal is at the receiver front end. However, the received signal excludes noise, interference, and other channel effects (e.g., multi-path fading and shadowing). Therefore, a high RSS does not always mean that the channel and signal quality are good. Instead, it gives an indication of whether a strong signal is present in the channel of interest. On the other hand, the CINR provides information on how strong the desired signal is compared to the interferer (or noise, or interference-plus-noise). In this study, an MN exploits a series of instantaneous CINR values that are used to derive the mean of the CINR using the moving mean method.

$$\mu CINR[k] = \frac{1}{N_m} \sum_{i=0}^{N_m} \mu CINR[k-i] \tag{1}$$

where, μCINR[k] is the mean of CINR during the k-th measurement and $N_m$ is the maximum number of measurement. We conduct a simulation experiment to investigate the performance of VoIP over 802.16e and the behavior of downlink CINR and MN's queue length using QualNet 4.5 [10]. We employ Mean Opinion Score (MOS) to assess the VoIP quality. MOS is defined in E-model standardized by ITU-T [11]. MOS provides a subjective numerical indication of the perceived quality of conversational voice call that ranges from 1 (lowest quality) to 5 (highest quality). MOS is determined based on

R-factor which mainly sum of delay and loss impairment [12]. Normally, MOS > 3.60 means an adequate VoIP call quality.

We set an urban environment in our simulation scenario by introducing a Rayleigh fading model to meet our scenario (see **Figure 1**(A)). In this simulation experiment, we employ VoIP codec of G.711, which are commonly used in the VoIP system where G.711 sends a packet of 160 bytes every 20 ms. The detail of simulation parameters describes in **Table 1**. Through simulation experiments, we aim to determine a certain level of threshold for downlink CINR and MN's queue length to meet the requirement of an adequate VoIP call quality.

**Figure 2** shows the relationship between Downlink MOS and CINR when an MN moves away from BS. Since the relationship starts to scatter when CINR below 50 dB due to the fluctuation of wireless link condition, we employ a local regression smoothing method (loess method) to grasp their trend and estimate the best fit of CINR level and MOS over the distance shown as a red curve. The red curve shows that CINR level above 26 dB satisfies an adequate VoIP call. Therefore, we can set the CINR level of 26 dB as a threshold for initiating HO in this study.

### 3.2.2. MN's Queue Length

In the 802.16e standard, BW allocation is thoroughly managed by a BS, in which the downlink BW is managed by a BW scheduler at the BS. On the other hand, uplink BW is allocated by the BS through the BW request and uplink grant procedures. In particular to uplink transmission, as shown in **Figure 3**, the BS allocates BW to the MNs for the purpose of making BW requests by providing a BW request contention slot in UL-MAP. Then, the MN transmits the BW requests in response to the allocated time slot. After the MN obtains an Uplink Grant, it can then transmit its data with the allocated uplink grants.

So, we can sum up the total end-to-end uplink delay ($T_{UL-ETE}$) as follows:

$$T_{UL-ETE} = T_{BWreq} + T_{Sch} + T_{queue} + T_{o} \quad (2)$$

where $T_{BWreq}$ and $T_{Sch}$ are BW request contention and BS scheduling delay, respectively. $T_{queue}$ represents the queuing delay in the transmission buffer. $T_{o}$ represents additional delays, e.g., transmission delays over wireless link and VoIP codec delay. This BW request contention and scheduling delay may becomes longer with the increase in the number of MNs in a serving BS, which contends the BW request slot in the UL-MAP.

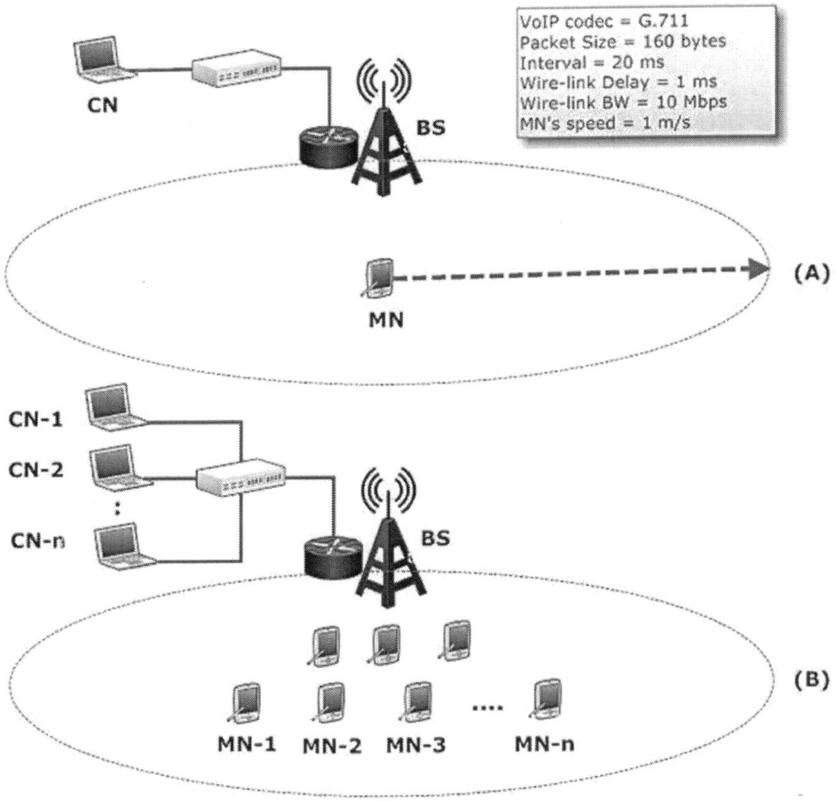


**Figure 1**. Simulation Scenarios.

**Table 1**. Simulation Parameter For 802.16e.

| | |
|---|---|
| Frequency Band | 2.4 GHz |
| Duplexing Mode | TDD |
| Channel Mode | 10 MHz |
| FFT Size (of OFDMA sub-carriers) | 1024 |
| Frame Duration | 5 ms |
| Transmit/Receive Time Gap (TTG) | 10 us |
| Receive/Transmit Time Gap (RTG) | 10 us |
| BS Tx Power | 30 dBm |
| MN Tx Power | 22 dBm |
| Fading Model | Rayleigh |

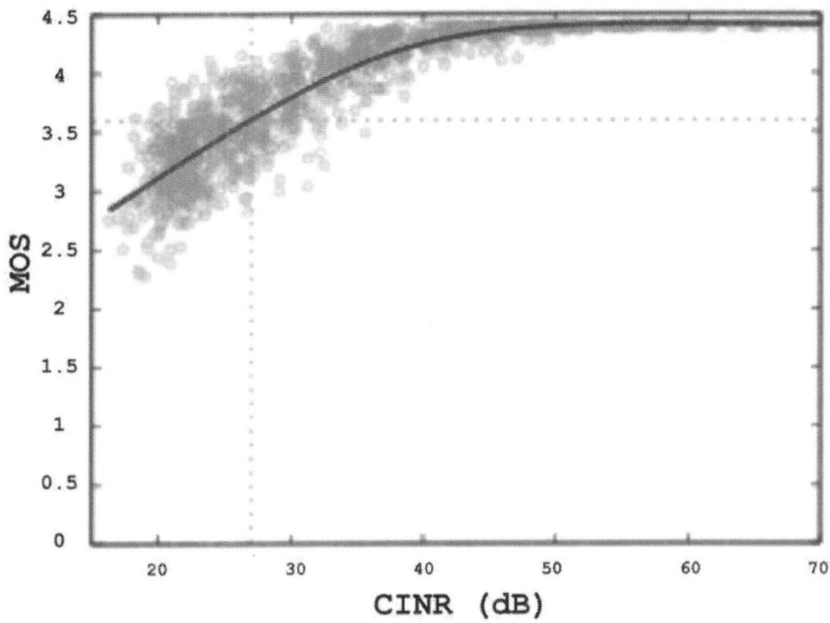


**Figure 2**. MOS vs. CINR.

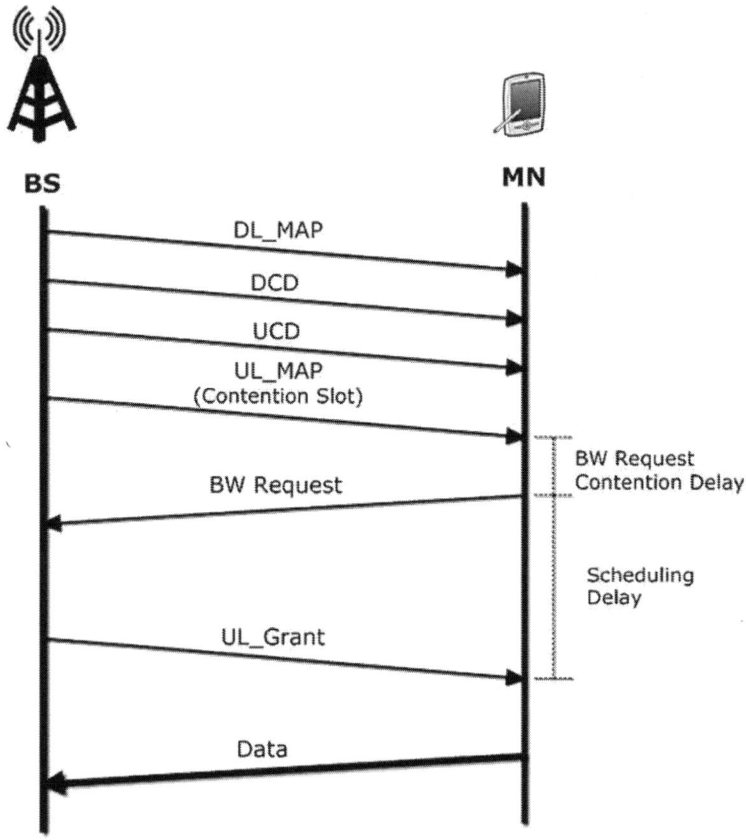


**Figure 3**. Uplink Transmission of 802.16e.

The increase of BW request contention and scheduler delay cause the increase of data on the queue length in the MN for transmission (MN's queue length), thereby leading to a degradation of uplink communication quality. We conduct a simulation experiment to investigate the relationship between the number of MNs establishing VoIP calls and MN's queue length. In this simulation scenario, we randomly locate from one to 30 MNs within a coverage area of the single BS. Each MN communicates with a Correspondent Node (CN) using VoIP. **Figure 1**(B) shows the simulation model where every MN moves randomly at a speed of 1 m/s within a coverage area of the single BS.

**Figure 4** shows the average uplink and downlink VoIP quality as well as the average of MN's queue length over the number of MNs in the single BS. The figure shows that uplink VoIP quality decreases as the number of MNs in the single BS increases. On the other hand, downlink VoIP quality is kept at an adequate VoIP quality even if the number of VoIP calls increases. Moreover, the MN's queue length increases as the number of MNs establishing VoIP calls in the single BS increases. From the figure, we can see that the bottleneck of the 802.16e network affects only flows from the MN to BS, i.e., uplink transmission. This bottleneck causes the uplink latency, which is mainly raised from the uplink BW request contention and scheduling delay. This uplink latency causes an increase of the MN's queue length, thereby leading to the degradation of uplink communication quality.

In terms of accommodation of MNs establishing VoIP call in the single BS, **Figure 5** shows that a single BS can simultaneously accommodate 20 MNs VoIP calls. In this simulation experiment, we aim to determine the threshold level of MN's queue length to meet the requirement of an adequate VoIP call quality condition in the 802.16e network. Therefore, we show the relationship among MN's queue length, uplink VoIP quality and uplink latency in **Figure 6**. From this figure, we can see that the G.711 VoIP application has a tolerable uplink end-to-end delay up to 260 ms.

Since we are interested in MN's queue length, we can see from **Figure 6** that MN's queue lengths of less than 12,000 bytes meet the requirement for adequate VoIP call quality. Therefore, we set the MN's queue length threshold of 12,000 bytes that can be a reference for initiating the HO.

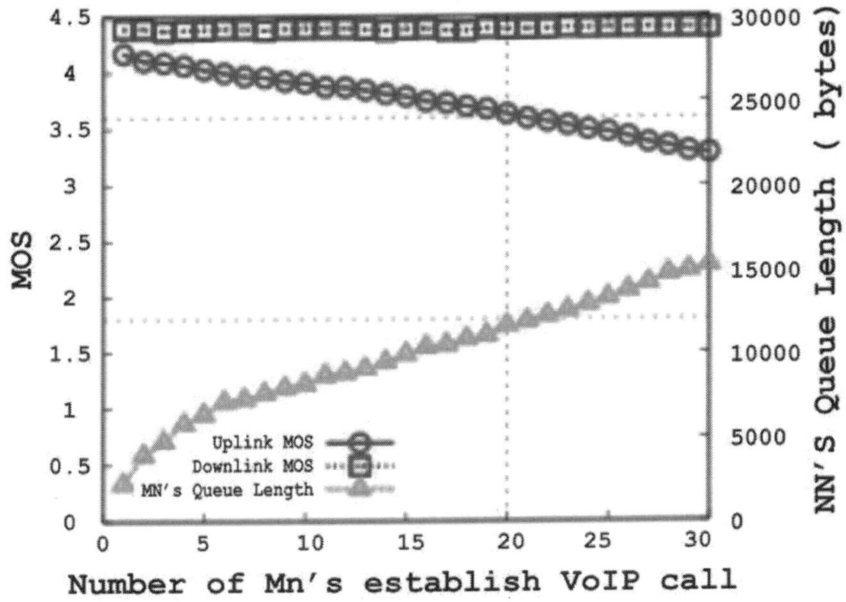


**Figure 4**. Relationship among MOS, MN's QL.

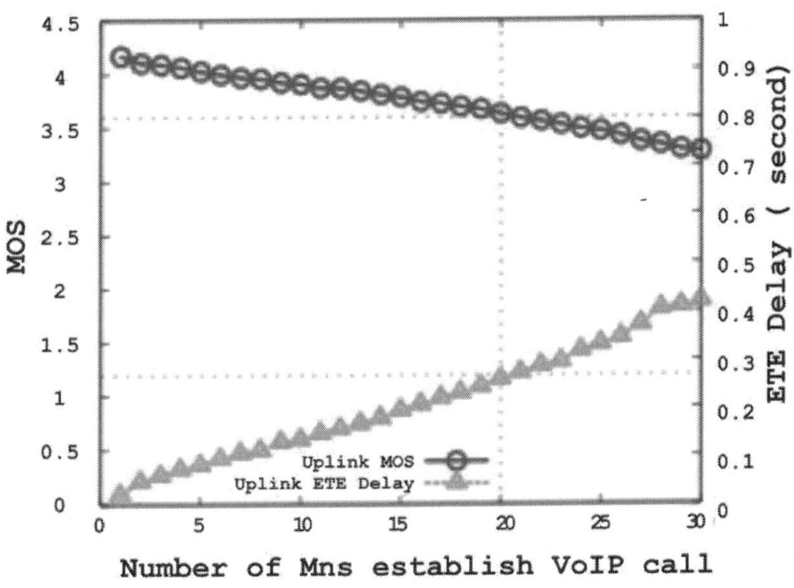


**Figure 5**. Relationship among MOS, ETE Delay and MN's QL.

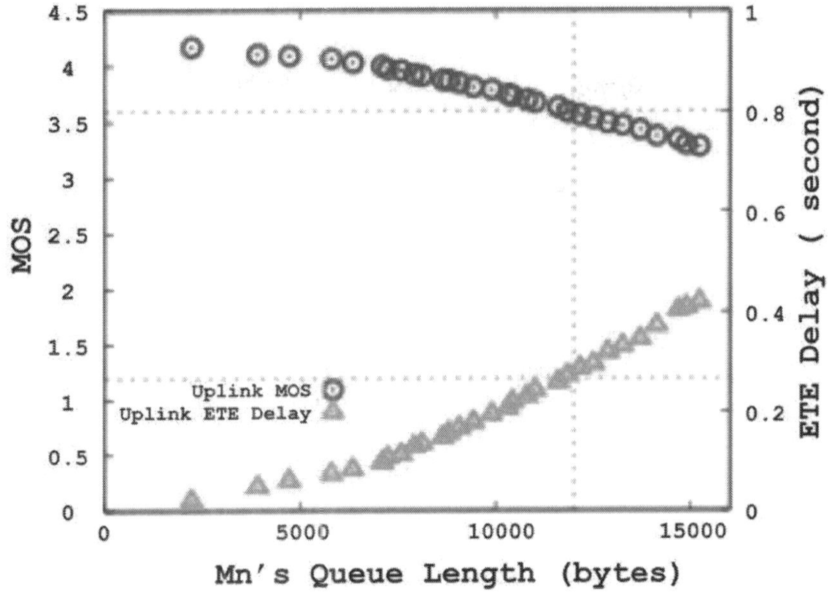


**Figure 6**. Relationship among MOS, ETE Delay and # of VoIP calls.

## 4. PROPOSED VHM ARCHITECTURE

We propose a VHM for intermingled 802.11g and 802.16e networks. Our proposed VHM is implemented on the transport layer of the MN and obtains effective HO triggers from a low layer using a cross-layer approach as shown in **Figure 7**.

### 4.1. VHM Features

In our study, we consider an MN has two IFs, i.e., 802.11g and 802.16e IFs, and each IF is assigned single IP address with different IP subnet. The MN has dual IFs connected to two point of attachment, i.e., BS and AP. The MN establishes VoIP call with the CN. One or both endpoints (CN or MN) can be multi-homed and multiaddressed but we here focus on that only MN is multihomed in this study. Utilizing multi-homing function can enhance the session preservation between MN and CN. The main purpose of utilizing the multi-homing function is that an MN may use its two IFs either alternatively or simultaneously in order to achieve seamless HO. An MN has a choice to select the best IF in order to preserve an on-going VoIP session while changing the point of attachment. The VHM selects the transmission mode, i.e., single/bi-casting, based on RTS retries and W-RTT of the 802.11g IF as well as CINR level and MN's queue length of the 802.16e IF as HO triggers. Our VHM is mainly focused on proper HO initiation according to wireless link and congestion of network condition.

## 4.2. Single/Bi-Casting Transmission Mode

As our proposed VHM employs multihoming similar to [8,9] in order to enhance the VoIP session preservation between MN and CN, the VHM must properly switches between single-casting and bi-casting modes in response to wireless network condition. Single-casting mode means that an MN communicates with a CN using only one IF. Bi-casting, on the other hand, means that an MN sends duplicated packets to a CN through two IFs for supporting soft-HO. Bi-casting is useful for avoiding packet loss and investigating both 802.11g and 802.16e IFs condition when an MN changes its IF during communication; hence, it achieves more reliable VoIP communication during HO.

Pseudo-code 1 shows an algorithm of switching to bicasting when an MN is located in an overlap area of 802.11g and 802.16e (AP and BS). Basically, our algorithm first examines the wireless link condition by observing RTS retry (rts_retry) of 802.11g IF and mean CINR level (cinr) of 802.16e IF. After that it examines the congestion state of wireless network by observing W-RTT (w-rtt) of 802.11g IF and MN's queue length (mn_queue) of 802.16e IF. For example, when the MN's initial transmission mode is single-casting via 802.11g IF, the VHM monitors the rts_retry to detect the wireless link quality. If the rts_retry exceeds threshold rts_thr (0.6) [6], the MN switches transmission mode to bi-casting; otherwise, it monitors w-rtt to detect the congestion of 802.11g network. If w-rtt exceeds threshold rtt_thr (200 ms) [6], the MN switches transmission mode to bi-casting; otherwise, it stays in single-casting via 802.11g IF. When the MN's initial transmission mode is single-casting via 802.16e IF, the MN monitors downlink cinr to detect the wireless link quality. If cinr is less than cinr_thr (26 dB), the MN switches transmission mode to bi-casting; otherwise, it monitors the mn_queue to detect the congestion of 802.16e network. If mn_queue exceeds the threshold queue_thr (12,000 bytes), the MN switches transmission mode to bi-casting; otherwise, it stays in single-casting via 802.16e IF.

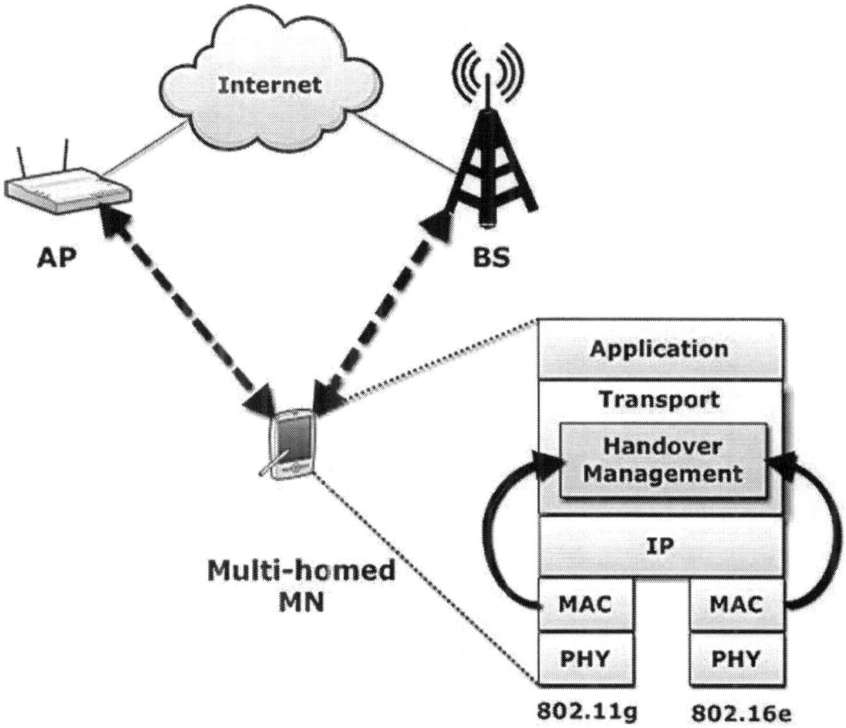


**Figure 7**. VHM Architecture.

When bi-casting is applied, the MN monitors all HO triggers from both IFs and compares them. As shown in Pseudo-code 2, an algorithm of switching to single-casting works as follows. An MN first monitors HO triggers for indicating wireless link condition and follow by congestion state of wireless network. First of all, an MN monitors and evaluates link state HO triggers, i.e., rts_rety and cinr, of 802.11g and 802.16e IFs, respectively. If both triggers still exceed and fall below the threshold for rts_thr (0.6) and cinr_thr (26 dB), respectively, the MN stays in bi-casting mode. If rts_retry exceeds the threshold while the cinr is still above threshold, the MN then switches to the single-casting via 802.16e IF; otherwise, it switches to single-casting via 802.11g IF.

When both triggers still exceed and fall below threshold for rts_thr and cinr_thr, respectively, the MN then monitors the congestion state HO triggers of both 802.11g and 802.16e IFs, i.e., w-rtt and mn_queue, respectively. If both congestion state HO triggers exceed the threshold, the MN stays in bi-casting mode. If w-rtt of 802.11g IF exceeds the threshold while mn_queue of 802.16e IF are below threshold, then, MN switches to the single-casting via 802.16e IF; otherwise, it switches to single-casting via 802.11g IF. When both triggers fall below threshold, the MN switches back to the initial transmission mode that is the same mode as when the MN established a connection in the first place. When both networks remain congested, the MN continues in bi-casting leading to more congestion in both networks. In this situation, we need to extend our proposed VHM with congestion control mechanism to deal with congestion of both networks. Since we focus on HO initiation and decision in this study, we leave this topic for future investigation.

# 5. SIMULATION EXPERIMENTS AND RESULTS

We conduct simulation experiments to investigate the effectiveness of our proposed VHM using QualNet 4.5 [10]. We evaluate our proposed VHM when degradation of wireless link occurs due to the movement of MN as well as when congestion of wireless network occurs due to increase of number of MNs sending VoIP traffic. In this simulation experiment, we employ VoIP codec of G.711, which is commonly used in VoIP systems where G.711 sends a packet of 160 bytes every 20 ms. The details of simulation parameters describes in Tables 1 and 2 for 802.16e and 802.11g, respectively.

```
BEGIN
        IF Single-casting if:802.11g
        READ Link State HO Triggers FROM if:802.11g
           IF rts_retry ≤ rts_thr THEN
                READ Congestion State HO Triggers FROM if:802.11g
                IF w-rtt ≤ rtt_thr THEN
                    Single-casting if:802.11g
                ELSE Bi-casting
                EN-
           DIF
           ELSE
              Bi-casting
           ENDIF
          ELSE IF Single-casting if:802.16e
            READ Link State HO Triggers FROM if:802.16e
            IF cinr ≥ cinr_thr THEN
      READ Congestion State HO Triggers FROM if:802.16e
                 IF mn_queue ≤ queue_thr THEN
                     Single-casting if:802.16e
                 ELSE
                       Bi-casting
                 ENDIF ELSE
                Bi-casting
            ENDIF ENDIF
        END
```

**Pseudo-code 1. Switching from single-to-bi-casting.**

Pseudo-code 1. Switching from single-to-bi-casting.

```
BEGIN
  IF Bi-casting
   READ Link State HO Triggers FROM if:802.11g && 802.16e
    IF rts_retry ≥ rts_thr && cinr <cinr_thr THEN
        Bi-casting
    ELSE IF rts_retry ≥ rts_thr && cinr ≥ cinr_thr THEN
      Single-casting if:802.16e
    ELSE IF rts_retry <rts_thr && cinr <cinr_thr THEN
      Single-casting if:802.11g
    ELSE
     READ Congestion State HO Triggers
     FROM if:802.11g&802.16e
      IF w-rtt ≥ rtt_thr && mn_queue ≥ queue_thr THEN
          Bi-casting
      ELSE IF w-rtt ≥ rtt_thr && mn_queue <queue_thr THEN
          Single-casting if:802.16e
      ELSE IF w-rtt < rtt_thr && mn_queue ≥ queue_thr THEN Single-casting
      if:802.11g
      ELSE Bi-casting
      ENDIF ENDIF ENDIF
     END
```

Pseudo-code 2. Switching from bi-to-bi/single-casting.

Our proposed VHM can be applied in a condition where an 802.16e network cell overlays an 802.11g network cell; hence, an MN located in the 802.11g network cell can associate with the 802.16e network. We assume that both IFs are in standby mode. Therefore, the MN does not need to perform network acquisition including network scanning and ranging for 802.11g and 802.16e, respectively. Therefore, we neglect the delay caused by network acquisition.

## 5.1. Performance Evaluation of HO Based on Link State HO Triggers

This simulation experiment aims to investigate whether our proposed VHM can initiate HO and decide the transmission mode according to link state HO triggers, i.e., RTS retry of 802.11g IF and CINR level of 802.16e IF. Figures 8(A) shows the simulation model to evaluate the proposed HO algorithms. An MN, which is establishing a VoIP call with a CN, moves from the 802.11g area to the 802.16e area at 1 m/s, and vice versa.

In terms of VoIP quality, we can see that our proposed method can obtain average MOS values of 4.06 (**Figure 9**(a)) and 4.286 (**Figure 9**(b)) for uplink and downlink, respectively, when an MN moves from 802.11g to 802.16e. From **Figure 9**(a), the average uplink MOS is a bit decreased compared to downlink MOS due to uplink latency of 802.16e, which is mainly raised from the uplink BW request contention and scheduling delays.

**Table 2**. Simulation Parameter for 802.11g.

| | |
|---|---|
| Supported Data Rate (Mbps) | 6, 9, 12, 18, 24, 36, 48, 54 |
| Tx Power (dBm) | 5.0, 5.1, 6.7, 7.9, 8.1, 9.3, 10.6, 10.1 |
| SIFS | 16 us |
| Slot Time | 9 us |
| CWmin, CWmax | 15, 1023 |

On the other hand, when an MN moves from 802.16e to 802.11g, average MOS values are 4.29 (**Figure 10**(a)) and 4.28 (**Figure 10**(b)) for uplink and downlink, respectively. MOS is decreased sometimes below 3.6 at simulation time of 200 seconds. This is mainly because the MN located in the edge of 802.11g network at 200 seconds where link condition is degraded and also effect of fading model introduced in the simulation scenario.

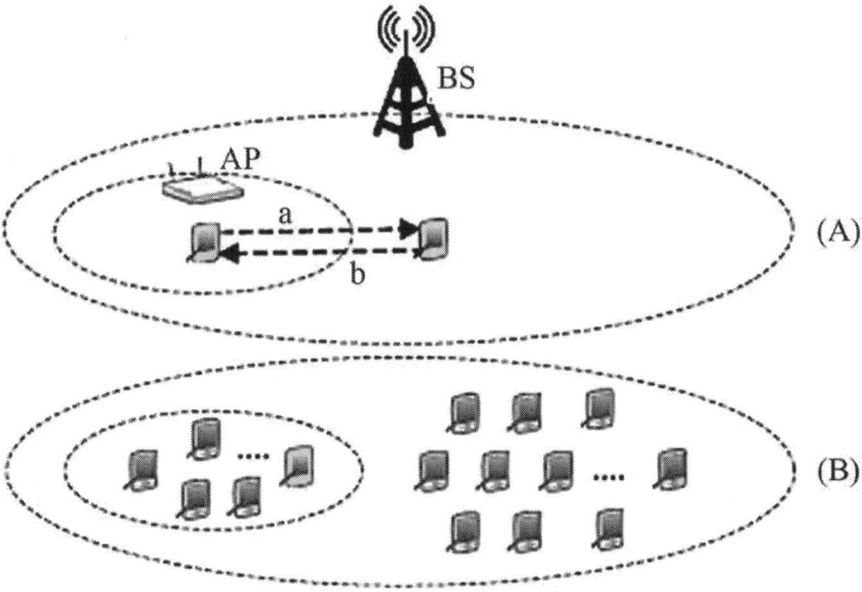


**Figure 8**. Simulation Scenarios for Evaluating Proposed VHM.

**Figure 11** shows the average MOS, as a function of distance between BS and AP. This figure shows that as an increase of distance between AP and BS, the average uplink and downlink MOS (when an MN moves from 802.11g to 802.16e and vise versa) are gradually decreased. This is mainly because an overlap area between 802.11g and 802.16e becomes less as the increase of distance between AP and BS.

Since our proposed VHM can be applied in a situation where a network cell of 802.16 overlays a network cell of 802.11g, lessen overlap area between two networks will affect the VoIP call quality. From **Figure 11**, the effecttive distance of 300 meters between AP and BS can be applied to our proposed VHM.

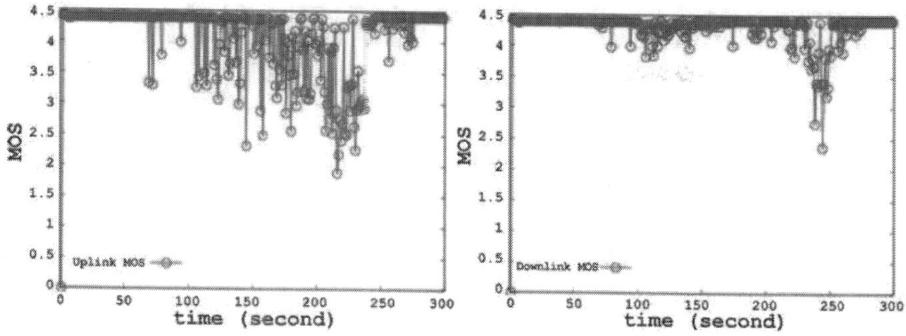


(a) Uplink MOS

(b) Downlink MOS

**Figure 9**. MN moves from 802.11g to 802.16e network.

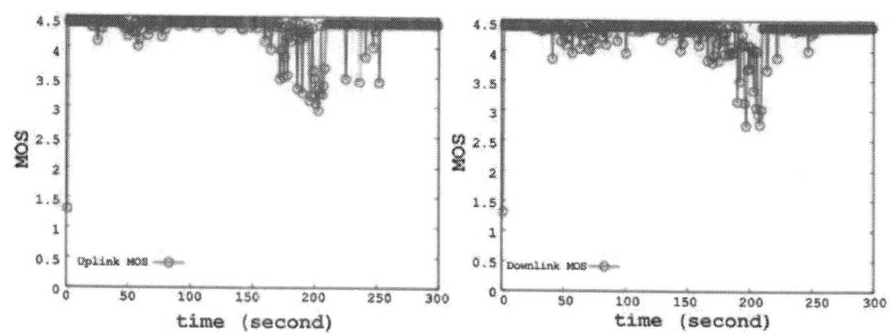


(a) Uplink MOS

(b) Downlink MOS

**Figure 10**. MN moves from 802.16e to 802.11g network.

## 5.2. Performance Evaluation of HO Based on Congestion State HO Triggers

This simulation experiment aims to investigate whether our proposed HO algorithm can initiate HO and decide the transmission mode according to congestion state HO triggers, i.e., W-RTT of 802.11g IF and MN's queue length of 802.16e IF. First, we conducted simulation experiments to evaluate the effectiveness of our proposed VHM in a congested 802.11g network. We consider the scenario depicted in **Figure 8**(B). At the beginning of the simulation, there are 13 MNs randomly distributed in the vicinity of an AP (i.e., between 0 and 50 m) where the access to 802.16e networks is granted.

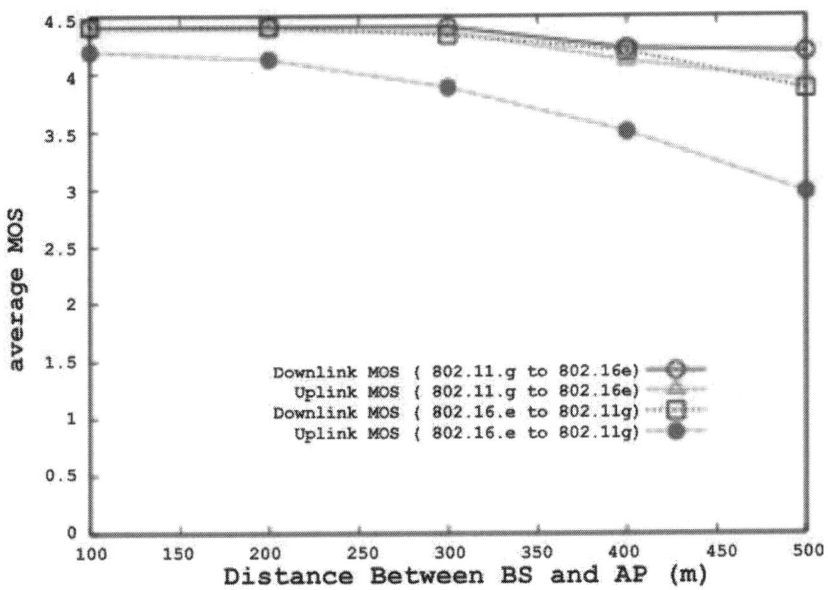


**Figure 11**. MOS vs. Distance between BS and AP.

There is one among many MNs employing the proposed HO algorithms and it initially uses the 802.11g AP to establish a VoIP call. Every 5 seconds, a new MN establishes a VoIP connection with its CN via the 802.11g network, so that the traffic in the 802.11g network gradually increases. Simulation results show that the average uplink and downlink MOS are 4.258 and 4.249, respectively (Figures 12(a) and (b)). From these figures, the MOS is decreased sometimes below 3.6 at simulation time of 50 seconds. This is mainly because the number of MN reaches to maximum acceptable number of MNs establishing VoIP call in single AP of 802.11g, i.e., 10 MNs, at the time of 50 seconds. After HO to 802.16e, uplink MOS is decreased sometimes at initial time because uplink latency due to BW request contention slot and scheduling at BS in 802.16e. Furthermore, we also evaluate the effectiveness of our proposed VHM in a congested 802.16e network.

At the beginning of the simulation, there are 30 MNs randomly distributed in the vicinity of AP and BS. We set 30 MNs in order to make a congested scenario in single BS (the acceptable number of MNs establishing VoIP call in the single BS of 802.16e is 20 MNs as shown in **Figure 4**). There is one among many MNs employing the proposed VHM algorithms and it initially uses the 802.16e to establish a VoIP call. Every 3 seconds, a new MN establishes a VoIP connection with its CN via the 802.16e network, so that the traffic in the 802.16e network gradually increases. The simulation results show that average uplink and downlink MOS are 3.882 and 4.336, respectively (Figures 13(a) and (b)). The uplink MOS suffers degradation many times from uplink latency due to BW request contention and scheduling delay.

## 6. CONCLUSIONS AND FUTURE WORK

In this paper, we have proposed VHM for VoIP over intermingled 802.11g and 802.16e networks. Our proposed VHM is implemented on the transport layer of the MN and obtains effective HO triggers from the lower layer using a cross-layer approach. The VHM exploits RTS retries and W-RTT of the 802.11g IF as well as CINR level and MN's queue length of the 802.16e IF as HO triggers. Our proposed method aimed to preserve VoIP quality during HO between networks with different IP subnets. Our simulation results show that our proposed VHM can preserve VoIP call quality during such HOs.

In this study, we focused on HO initiation and decision to switch the transmission mode, i.e., single/bi-casting according to HO triggers obtained from 802.11g and 802.16e network. In the future, we need to extend our proposed VHM with congestion control mechanism to deal with the condition where both 802.11g and 802.16e networks are congested.

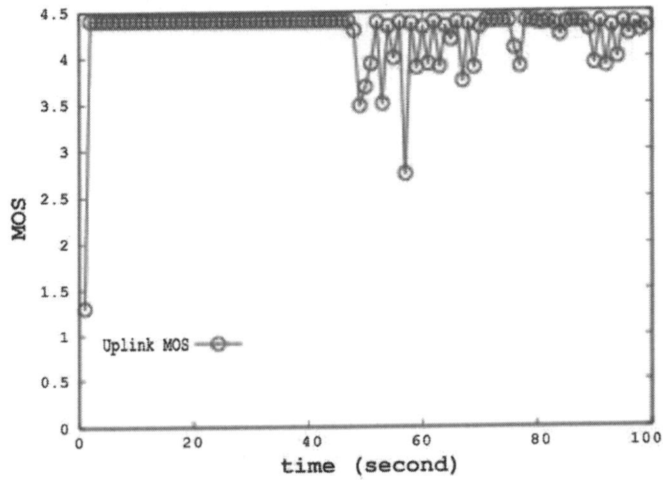


(a) Uplink MOS

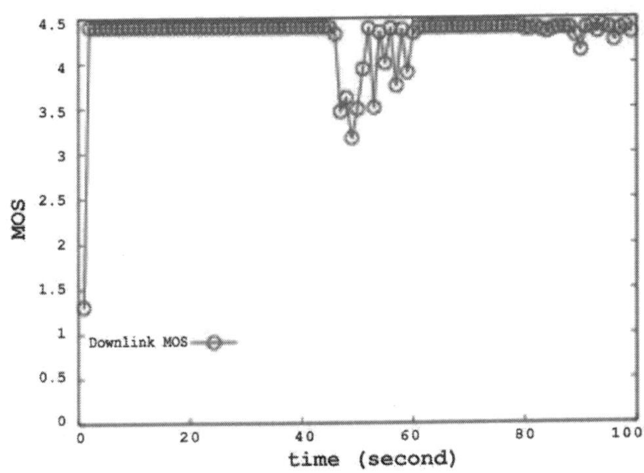


(b) Downlink MOS

**Figure 12**. 802.16e is congested, MN handovers to 802.11g.

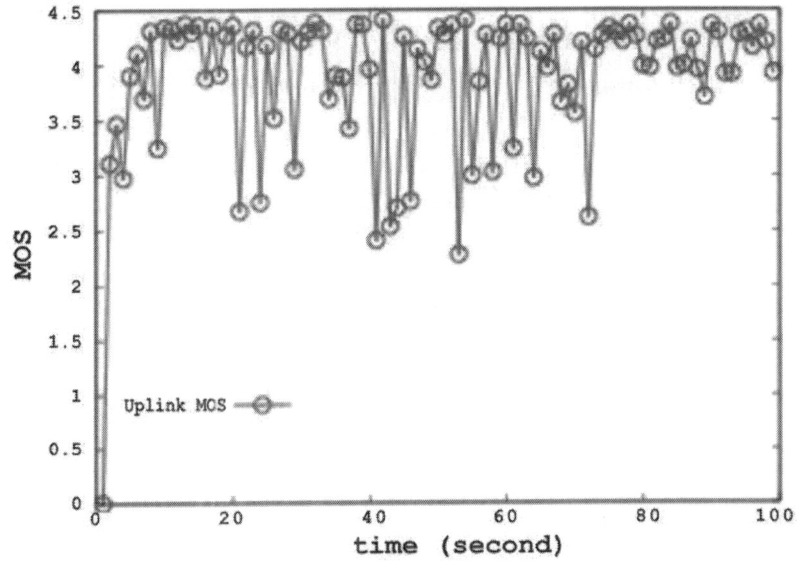


(a) Uplink MOS

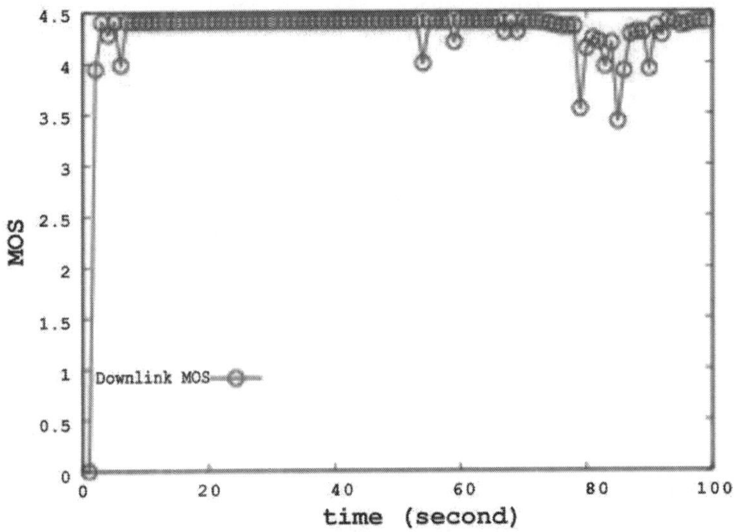


(b) Downlink MOS

**Figure 13**. 802.11g is congested, MN handovers to 802.16e.

# REFERENCES


1. A. H. Zahran and B. Liang, "Performance Evaluation Framework for Vertical Handoff Algorithms in Heteregeneous Networks," Proceedings of IEEE International Conference on Communications (ICC'05), Seoul, 16-20 May 2005, pp. 173-178.

2. S. Mohanty and I. F. Akyilidiz, "A Cross-Layer (Layer 2 + 3) Handoff Management Protocol for Next Generation Wireless Systems," IEEE Transactions on Mobile Computing, Vol. 5, No. 10, 2006, pp. 1347-1360.

3. H. Y. Liu, et al., "A Smart Triggering Scheme to Reduce Service Interruption during Heterogeneous Handovers," Proceedings of International Conference on Dependable Systems and Networks, Boston, 24-27 June2008.

4. B.-J. Chang and J. F. Chen, "Cross-Layer-Based Adaptive Vertical Handoff with Predictive RSS in Heterogeneous Wireless Networks," IEEE Transaction on Vehicular Technology, Vol. 57, No. 6, 2008, pp. 3679-3692.

5. Z. Dai, et al., "Vertical Handover Criteria and Algorithm in IEEE 802.11 and 802.16 Hybrid Networks," Proceedingds of IEEE International Conference on Communication (ICC2008), Beijing, 19-23 May 2008, pp. 2480- 2484.

6. A. S. Rashid, et al., "WiFi/WiMAX Heterogeneous Seamless Handover," Proceedings of 3rd International Conference on Broadband Communication, Pretoria, 23-26 November 2008, pp. 169-174.

7. V. Gupta, et al., "IEEE 802.21 Tutorial," 2006. http://www.ieee802.org/21/Tutorials/802%2021-IEEE-Tutorial.ppt

8. M. Niswar, et al., "Handover Management for VoWLAN Based on Estimation of AP Queue Length and Frame Retries," IEICE

Transaction on Information and System, Vol. E92-D, No. 10, 2009, pp. 1847-1856.

9. S. Kashihara and Y. Oie, "Handover Management Based on the Number of Data Frame Retransmissions for VoWLAN," Elsevier Computer Communications, Vol. 30, No. 17, 2007, pp. 3257-3269.

10. "Scalable Network Technology." http://www.scalablenetworks.com/

11. ITU-T Recommendation G.107, "The E-Model: A Computational Model for Use in Transmission Planning," 2005.

12. Cole, "Voice over IP Performance Monitoring," ACM SIGCOMM Computer Communication Review, Vol. 31, No. 2, 2001, pp. 9-24.

2

# A STUDY OF ARTIFICIAL SPEECH QUALITY ASSESSORS OF VOIP CALLS SUBJECT TO LIMITED BURSTY PACKET LOSSES

**Sofiene Jelassi and Gerardo Rubino**

INRIA Rennes - Bretagne Atlantique, Rennes, France

## ABSTRACT

A revolutionary feature of emerging media services over the Internet is their ability to account for human perception during service delivery processes, which surely increases their popularity and incomes. In such a situation, it is necessary to understand the users' perception, what should obviously be done using standardized subjective experiences. However, it is also important to develop artificial quality assessors that enable to automatically quantify the perceived quality. This efficiently helps performing optimal network and service management at the core and edges of the delivery systems. In our article, we explore the behavior rating of new emerging artificial speech quality assessors of VoIP calls subject to moderately bursty packet loss processes. The examined Speech Quality Assessment (SQA) algorithms are able to estimate speech quality of live VoIP calls at run-time using control information extracted from header content of received packets. They are especially designed to be sensitive to packet loss burstiness. The performance evaluation study is performed using a dedicated set-up software-based SQA framework. It offers a specialized packet killer and includes the implementation of four SQA algorithms. A speech quality database, which covers a wide range of bursty packet loss conditions, has been created and then thoroughly analyzed. Our main findings are the following: (1) all examined automatic bursty-loss aware speech quality assessors achieve a satisfactory correlation under upper (> 20%) and lower (< 10%) ranges of packet loss processes; (2) they exhibit a clear weakness to assess speech quality under a moderated packet loss process; (3) the accuracy of sequence-by-sequence basis of examined SQA algorithms should be addressed in detail for further precision.

## KEYWORDS

VoIP; QoE; Artificial speech quality assessors; Bursty packet losses

## 1. INTRODUCTION

Early telecommunication networks were engineered in such a way that enables offering a steady perceived quality of delivered services during a media session. This goal is achieved through the reservation of resources needed before launching services' delivery processes. Telecoms operators are impelled to select and install suitable transmission mediums and equipment that guarantee a standardized perceived quality for their customers independently of their geographical location and service delivery context. In such a situation, a client request is solely admitted if there are sufficient resources to accommodate it in the transport network. However, the introduction of 2G cellular telecom systems that deliver services to moving customers induces difficulties to conquer the challenge of keeping a time-constant perceived quality. The principal factors entailing perceived quality fluctuation are handovers among access points and vulnerability of wireless channels to unpredictable interferences and obstacles. It is worth to note here that keeping a steady perceived quality over a mobile telecom system is achievable, but the remedies are unreasonably expensive and impracticable for telecom operators. In reality, mobile customers are more tolerant and tend to accept fluctuations in the perceived quality during a media session given their awareness regarding mobile network features. The integration of delay sensitive telecom services over the best effort IP networks obviously emphasizes the fluctuation of perceived quality of delivered services.

There are a wide range of vital network-related operations where the accurate assessment of time-varying perceived quality is desirable and helpful [1, 2]. A reliable measure of perceived quality can be beneficial before, during, and after service delivery. The offline usages of perceived quality measurement include network planning, optimization, and marketing. The online usages of perceived quality measurement include networks and services management, monitoring, and diagnosis. This ultimately indicates that the use of perceived quality help decision makers to select choices that maximize profitability while maintaining an optimal user's satisfaction. Under the scope of this work, we explore the accurate estimation of perceived listening quality of PC-to-PC and PC-to-PSTN phone calls, denoted often as VoIP (Voice over IP), that currently live in their blossoming period.

A wide range of factors can affect the perceived quality of VoIP services, such as coding scheme, packet loss, noises, network delay and its variation, echoes, and handovers. Recent studies reveal that packet loss constitutes the principal source of perceived quality degradation of VoIP calls [1, 3]. The negative effect of missing packets is more disturbing especially when packets are removed in bursts, i.e., multiple media units are consecutively dropped from the original media stream. As a rule of thumb, the higher the loss 'burstiness degree', the greater the quality degradation. Unlike independent packet losses, missing media chunks under bursty packet loss processes exhibit high temporal dependency. This means that the probability of missing a given packet is much higher when the previous ones have been dropped. Figure 1a presents a packet loss pattern with independent packet losses. As we can observe, isolated and temporally-independent loss instances[a], denoted sometimes as loss islands, are introduced in the rendered stream. Figure 1b presents packet loss patterns following heavy bursty packet loss processes. Here, loss instances are temporally closed and may comprise multiple packets. A particular scenario of bursty packet loss processes is when isolated missing chunks are dropped with high frequency (see Figure 1c). This is referred to as sparse bursty packet losses. From users' perspective, each packet loss pattern generates a distinct perceived quality [3]. Therefore, the accurate measure of perceived quality needs to consider the prevailing packet loss pattern.

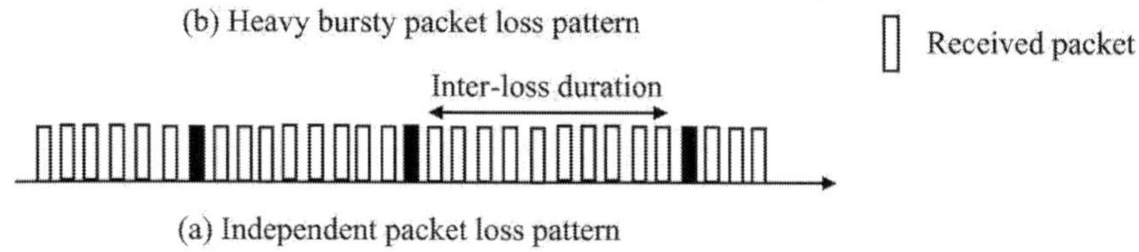


**Figure 1. Examples of independent, bursty, and sparse bursty packet losses. (a)** Independent packet loss pattern. **(b)** Heavy bursty packet loss pattern. **(c)** Sparse bursty packet loss pattern.

Basically, rather than the packet loss pattern itself, theoretical and representative models that capture the relevant features of packet loss processes are used for the estimation of the perceived quality for efficiency purposes. The characterization parameters are extracted from packet loss models that are calibrated at run-time using efficient packet-loss driven counting algorithms. Next, the effect of prevailing packet loss patterns can be judged using parametric assessment quality models built a priori. Typically, temporally-dependent packet loss processes are modeled using a simple, yet accurate 2-state discrete-time Markov chain, referred to as the Gilbert model, which has been well studied in the literature [3]. In a few words, Gilbert model has NO-LOSS and LOSS states that, respectively, represent successful and failing packet delivery operation. The Gilbert model is wholly characterized by the Packet Loss Ratio (PLR) and the Mean Burst Loss Size (MBLS) [4]. Typically, the higher the value of MBLS, the greater the burstiness of the loss process. For the sake of a more subtle characterization of packet loss processes, Clark [5] proposed a dedicated packet loss model that discriminates between isolated and bursty loss instances. The author defined adequate rules to classify loss instances either in isolated or bursty state and developed an efficient packet loss driven algorithm that enables to calibrate his enriched model at run-time. 'Appendix' section gives a survey about models of packet loss processes over VoIP networks.

This article explores the effectiveness of four single-ended bursty-loss aware Speech Quality Assessment (SQA) algorithms to evaluate the perceived quality of VoIP calls subject to distinct and limited bursty packet loss processes. To do that, a dedicated SQA framework has been set-up and a suitable SQA database has been built. It is crucial to note here that the perceived quality is automatically estimated using the double-sided signal-layer speech quality assessor defined in the ITU-T Rec. P.862, denoted as Perceived Evaluation of Speech Quality (PESQ), recognized by its accuracy to estimate subjective scores under a wide range of circumstances. The limitations of ITU-T PESQ have been considered in the design phase of the conducted empirical experiences, reducing its known defective behavior under 'generalized' bursty-packet loss processes (see below). To enhance measures' faithfulness, data filtering procedures have been applied on gathered raw ITU-T PESQ scores that involve outliers' detection and removal, coupled with the computation of the average scores among re-iterated experiences of each considered condition. Moreover, our study investigates the perceived effect of Comfort Noise (CN) and frequency bandwidth changeover required for speech material preparation. A statistical analysis has been conducted that enables drawing some conclusions about the rating behavior of existing bursty-loss aware SQA algorithms. As such, a set of potential clues for a better and consistent judgment accuracy of VoIP calls at run-time are identified and summarized.

The following sections are organized as follows. 'A review of SQA algorithms sensitive to packet loss burstiness' section reviews the four examined SQA algorithms that subsume packet loss burstiness. 'Set-up SQA framework and measurement strategy' section presents our set-up speech quality framework and measurement strategy. 'Speech material preparation and configuration parameters selection' section describes and discusses speech material preparation processes. A performance evaluation analysis is presented in 'Performance analysis of bursty-loss aware SQA algorithms' section. Concluding remarks and perspectives are given in 'Concluding remarks and perspectives' section.

## 2. A REVIEW OF SQA ALGORITHMS SENSITIVE TO PACKET LOSS BURSTINESS

The next sections introduce four SQA algorithms that will be thoroughly evaluated later. The shared feature of examined artificial speech quality assessors resides in their sensitivity to the different degrees of packet loss burstiness sustained by a VoIP packet stream.

## 2.1 VQmon: Voice Quality monitoring

VQmon is an early SQA algorithm intended to evaluate VoIP calls delivered over communication channels offering a time-varying quality [5]. Precisely, the delivery channel status alternates between Good and Bad states that refer to periods of time where packet loss ratio is low and high, respectively. In such a context, it is obvious to differentiate between intermediate and overall rating factors, denoted, respectively, hereafter as $R_I$ and $R$, that vary between 0 (Poor Quality) and 100 (Toll Quality). Specifically, the rating factor $R_I$ quantifies the perceived quality at the end of an independent short interval of duration 2 to 5 s. The rating factor $R$ quantifies the perceived quality at the end of a presented speech sequence. Moreover, earlier listening subjective tests of time-varying speech quality revealed that improvement (resp. degradation) of speech quality upon a transition from high to low (resp. low to high) loss periods is detected by subjects with some delay [6]. As such, immediate switching between plateaus $R_I$ values was found unnatural. This observation leads to define the notion of the perceptual instantaneous rating factor, $R_P$, which denotes the satisfaction degree at an arbitrary instant during the presentation. Figure 2 illustrates the evolution of $R_I$ (dashed line) and $R_P$(solid line) as function of time and channel state during a presented speech sequence.

VQmon models the evolution of the perceptual instantaneous rating factor, $R_p$, at the transition from high to low loss periods using an exponential decay, where the rapidity of the descent is calibrated according to subjective results [6]. Formally speaking, VQmon uses functions (1) and (2) to capture users' rating behavior at the transition from Good to Bad state, and conversely.

$$R_P(x) = R_I(t_k) + [R_P(t_{k-1}) - R_I(t_k)] \cdot e^{-(x-t_{k-1})/\tau_1}, \quad (1)$$

$$R_P(y) = R_I(t_{k+1}) - [R_I(t_{k+1}) - R_P(t_k)] \cdot e^{-(y-t_k)/\tau_2}, \quad (2)$$

where $t_i$ is the switching instant from ($i$-1)th to $i$ th segment, $R_I(t_i)$ refers to the intermediate rating factor estimated during the interval $[t_i, t_{i+1}]$, $R_P(t_i)$ refers to the perceptual instantaneous rating factor estimated at the instant $t_i$. The time variable $x$ refers to the prevailing instant in the speech presentation. The time constants $\tau_1$ and $\tau_2$ are used to calibrate the rapidity of the exponential decay at the transition from Good to Bad state, and conversely[b]. In the scope of VQmon, the value of $R_I$ is automatically estimated based on a directory of empirical subjective results that holds a mapping between the average PLR values and subjective rating factors.

At the end of a listened sequence, VQmon extracts packet loss characterization metrics, e.g., interval durations and their corresponding Good/Bad status and features, from a 4-state chain calibrated at run-time (see 'Appendix' section for further details). These control data are used to calculate the overall rating factor as follows, the built perceptual instantaneous rating function $R_P$ over a given Good and the next adjacent Bad segment is integrated over time. Then, the obtained value is divided by the interval duration. The resulting rating factor is referred to as average rating factor, $R_i$ (av), where the index $i$ represents the number of $i$ th good/bad segment (see Figure 2).

The limited subjective tests conducted by Clark showed that most of the time VQmon predicts with acceptable accuracy subjective rating of time-varying speech quality. In our opinion, the key shortcoming of VQmon resides in its incapability to accurately estimate $R_I$ value under bursty packet loss behavior. In fact, VQmon quantifies the effect of a bursty packet loss process solely using PLR value. As such, there is no subtle characterization and specification of the burstiness of the packet loss processes. This could lead to a wrong judgment of perceived quality because it has been subjectively observed that two distinct bursty packet loss patterns with identical PLR may lead to an obvious difference in the perceived quality [7]. Moreover, the rapidity of the exponential decay/growing is hold static independently of the duration of preceding Good or Bad state and the magnitude variation of previous and current packet loss ratios.

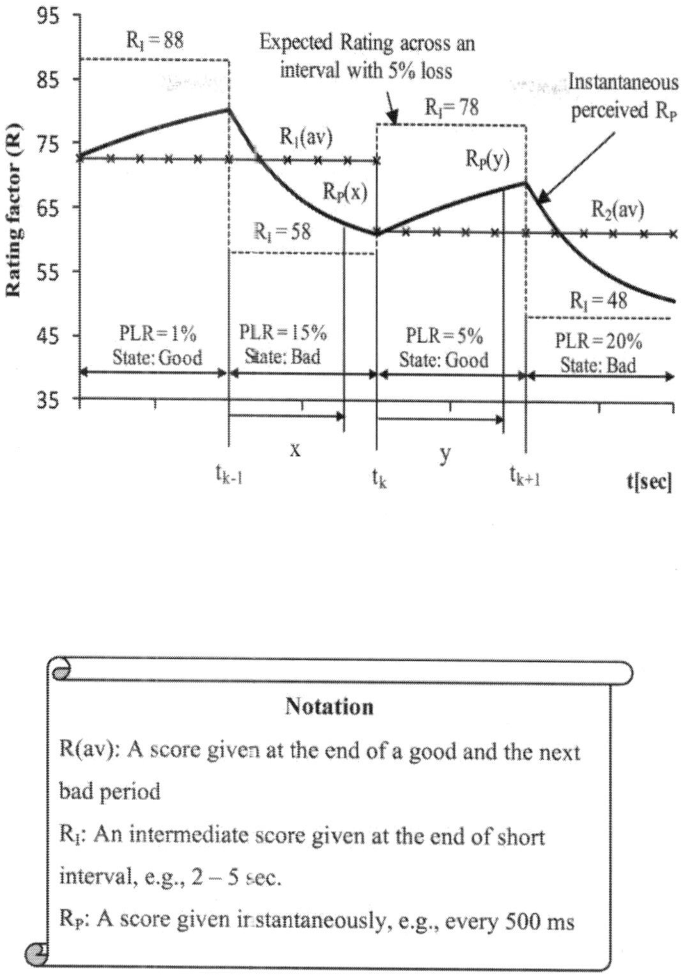


**Figure 2.** Modeling of intermediate and perceived quality behavior rating.

## 2.2 E-Model

The ITU-T defines in Rec. G.107 a computational model for use in planning of telephone networks, known as E-Model [8]. Briefly, the E-Model combines a set of characterization metrics of the transport system and provides as output a rating factor, *R*, that quantifies the users' satisfaction. The ultimate objective of E-Model consists of giving a synthesized overview regarding the perceived quality delivered over a given telecom infrastructure. It has been subsequently extended to consider packet-based telephone networks and to operate as a single-ended speech quality assessor [9]. The original release of the E-Model solely considers the negative perceived effect of independently removed voice packets. It has been recently evolved to account for bursty packet loss processes characterized using two newly defined parameters [8]. The first metric, denoted as BurstR, is defined as the ratio between the undergone average number of successive missing packets and the expected average number of successive missing packets under independent packet losses[c]. The second metric, denoted as $B_{pl}$, is a constant defined to consider the robustness of a given couple of CODEC and Packet Loss Concealment (PLC) algorithm to deal with bursty packet loss processes. The value of$B_{pl}$ is derived a priori for each CODEC and PLC algorithm using subjective tests and a comprehensive regression analysis [3].

Both BurstR and $B_{pl}$ metrics are used in the calculation of the effective equipment impairment factor,$I_{e,\ eff}$, that basically quantifies distortions caused by the coding scheme and the packet loss processes. The diagram given in Figure 3 summarizes the methodology followed to compute the value of $I_{e,\ eff}$ under a given configuration. As we can see, a real coefficient $0 \leq W \leq 1$ is calculated as a function of the variables PLR and BurstR, and the constant $B_{pl}$ (see Figure 3). The distortions caused by packet losses under a given coding scheme are captured by an impairment factor denoted as $I_{e,\ loss}$. It is obtained through the multiplication of the inherent achievable quality, ($95 - I_{e,\ codec}$), and $W$. Finally, the value of $I_{e,\ eff}$ is obtained by adding distortions caused by the coding scheme under no-loss condition, $I_{e,\ codec}$, and those caused by packet losses, $I_{e,\ loss}$.

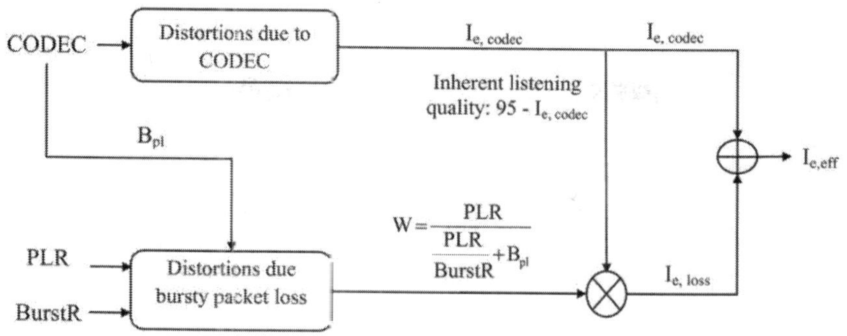


**Figure 3.** The measurement of quality degradations caused by coding scheme and bursty packet loss processes.

For the sake of planning, one can assume that sustained bursty packet loss processes exactly follow a Gilbert model that is wholly characterized using the PLR and CLP[d]. In such a case, the value of MBLS required to calculate BurstR is equal to 1/(1 - CLP). The curves plotted in Figure 4a show that bursty packet loss processes (i.e., where BurstR > 1) produce higher quality degradations than with independent losses (BurstR = 1) for an identical PLR. This is clearly observed especially for PLR greater than 4%. Figure 4b shows the quality degradation under different packet loss burstiness conditions. Basically, for a given PLR, the higher the packet loss burstiness, the greater the observed quality degradation.

The previously defined metrics for the characterization of packet loss burstiness explicitly (resp. implicitly) consider the nominal average length of sustained loss instances (resp. inter-loss durations). This could raise a biased quality rating factor because the subtle details of packet loss patterns are definitely ignored. The next presented speech quality assessors will consider this concern in a more careful fashion.

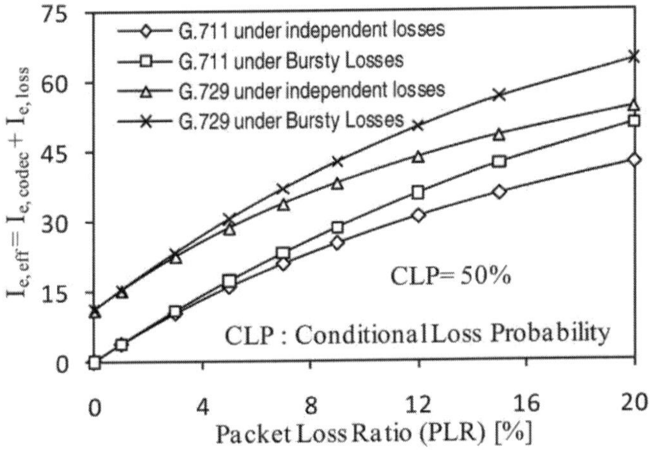


(a)

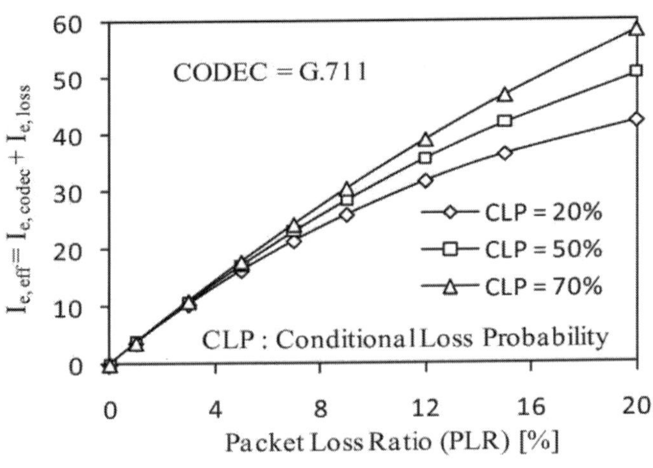


(b)

**Figure 4. The quality degradation as a function of packet loss burstiness. (a)** Quality degradation under independent and bursty packet loss processes. **(b)** Quality degradation as function of PLR and packet loss burstiness.

## 2.3 Genome

As outlined before, the previously described speech quality assessors capture the burstiness of packet loss processes using global characterization parameters. Hence, the concrete packet loss pattern is poorly considered in the estimation of the listening perceived quality. To overcome this shortage, Roychoudhuri and Al-Shaer [10] proposed a subtle grained speech quality assessor, denoted as Genome, that more accurately considers the pattern of dropped voice packets. To do that, a set of 'base' quality estimate models which quantify the perceived quality entailed by the application of a periodic packet loss processes[e] were developed, following a simple logarithmic regression analysis. The base quality estimate models are parameterized using the inter-loss gap and burst loss sizes. Specifically, for a packet loss run equal to 1, 2, 3, or 4 packets, a dedicated base quality estimate model, which has as input parameters the inter-loss gap size, has been built.

At run-time, Genome probes and records the effective experienced inter-loss gap and the following burst loss size. At the end of a monitoring period, the overall listening quality is computed as the weighted average of the 'base' quality score of each pair, where the weights are calculated as a function of the inter-loss gap durations (see Figure 5). Notice that the combination formula of Genome implies that the larger the inter-loss gap size of a given pair, the greater the influence on the overall perceived quality. Moreover, a high frequency of a given pair entails more impact on the overall perceived quality. These statistical properties of Genome can result in a biased behavior rating. Moreover, the fine granularity of Genome considerably disables its ability to consider the context in which a given loss instance happens. This perhaps explains why the authors confined the performance evaluation of Genome to independently dropped speech packets.

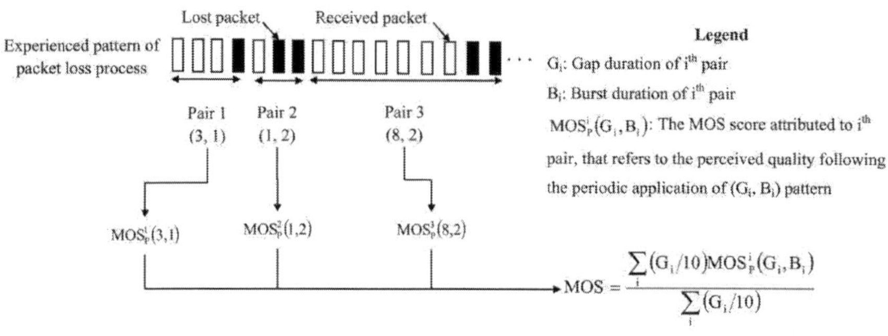


**Figure 5.** SQA methodology followed by Genome.

## 2.4 Q-Model

It is recognized that existing quality models are sufficiently accurate to estimate listening perceived quality of speech sequences subject to independent packet losses using PLR metric. This fact was the stimulus for the development of the speech quality assessor Q-Model reported in [11]. In such a case, the concern consists of finding the optimal PLR value of the independent packet losses that generates the equivalent perceived quality of a sustained bursty packet loss pattern. The curves plotted in Figure 6 illustrate the logic behind the equivalent perceived quality. The dashed line refers to quality degradation caused by independent packet losses. The other two solid lines represent quality degradation under two different bursty packet loss processes. As expected, independent packet losses produce the smallest degradation of perceived quality. The example given in Figure 6shows that for a given PLR value, $P_M$, different levels of quality degradation are observed according to the burstiness of the packet loss processes. For a measured PLR value equal to $P_M$, the independent packet losses processes that generate the equivalent perceived quality of first and second bursty packet loss processes are characterized by PLR values equal to $P_{E1}$ and $P_{E2}$, respectively.

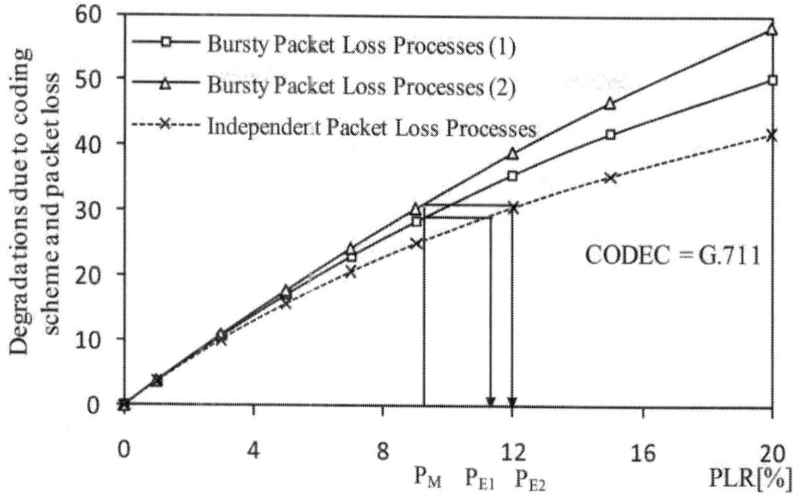


**Figure 6.** Equivalence between independent and bursty packet loss processes in term of quality degradation.

The Q-Model uses the following equation to determine the PLR of independent packet losses that produces the equivalent perceived quality of an observed bursty packet loss pattern:

$$PLR_E = PLR_M + \sum_{n=0}^{N-1} \alpha_n B_n, \tag{3}$$

where, $PLR_M$ refers to the measured packet loss ratio, $N$ is the total number of packets, and $\alpha_n$ is the weighting coefficient that has been derived following empirical trials[f][11]. The variable $B_n$ quantifies the local packet loss burstiness that is only calculated if the $n$ th packet is missing, otherwise it is set to 0. The value of $B_n$ is obtained according to the prevailing distances that separate the current missing packet, $n$, and previous ones along a monitoring window[g] with a fixed length equal to $N_{max}$. Basically, the larger the distance between successive missing packets, the lower the value of $B_n$. After an empirical study, the authors proposed the following equations to compute $B_n$:

$$B_{n,\mathrm{ed}} = \sum_{i=1}^{N_{\max}} \frac{P_{n-i}}{2^{i-1}} \quad \text{and} \quad B_{n,\mathrm{ld}} = \sum_{i=1}^{N_{\max}} \frac{P_{n-i}}{i}, \qquad (4)$$

where $B_{n,\ \mathrm{ed}}$(resp. $B_{n,\ \mathrm{ld}}$) refers to the exponential (resp. linear) dependency measurement strategy. The value of $B_{n,\ \mathrm{ed}}$(resp. $B_{n,\ \mathrm{ld}}$) geometrically (resp. linearly) decreases as the distance between two missing packets increases.

# 3. SET-UP SQA FRAMEWORK AND MEASUREMENT STRATEGY

The diagram given in Figure 7 illustrates the main building blocks of our set-up SQA framework. In short, a lossless stream of voice packets is created for each treated speech sequence following a specific encoding scheme and packetization strategy. The lossless packet stream goes through a packet killer that removes packets following a Gilbert model calibrated using PLR and MBLS values (see Figure 7). A degraded speech sequence is created according to the dictated pattern of missing packets. The lossless speech sequence is compared at the signal level to the lossy one using the SQA algorithm defined in ITU-T Rec. P.862, a.k.a PESQ [12]. PESQ is well-recognized by its good correlation and accuracy to estimate subjective LQ (Listening Quality) scores [12]. Note that this methodology has been advocated and followed by several researchers to avoid time, space, and budget costly subjective tests [1]. The quality scores calculated by PESQ are given on the MOS scale, i.e., between 1 (Poor Quality) and 5 (Excellent). However, apart Genome, the remaining examined SQA algorithms produce quality scores on the *R* scale. That is why, PESQ scores are mapped to the corresponding *R* factor using a standardized function given in ITU-T Rec. G.108 (see Figure 7). As we can note in Figure 7, we use the term 'measured' scores to refer to values calculated using PESQ algorithm and 'estimated' scores to refer to values returned by examined speech quality assessors. This terminology has been adopted since PESQ algorithm subtly models the processing behavior of the human auditory system in temporal and frequency domains. As such, PESQ scores can be seen as virtually measured scores that replace to a certain extent subjectively measured values.

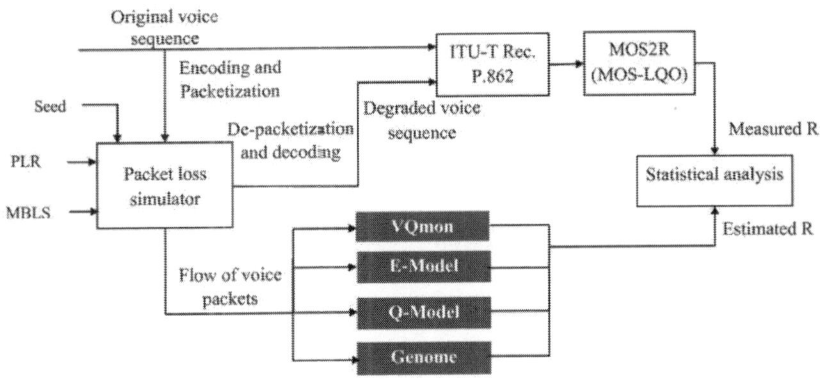


**Figure 7.** Diagram of developed SQA framework for the evaluation of VoIP calls.

It is worth to note here that typical VoIP applications install packet loss protection mechanisms at application and/or CODEC levels such as Forward Error Correction (FEC) or interleaving, in order to recover dropped voice packets in the network. Moreover, an adaptive de-jittering buffer is usually deployed that enables smartly reducing losses caused by late arrivals. Both, packet loss recovery schemes and de-jittering buffer policies are implicitly considered in our context because the considered packet loss pattern is monitored at the input of the speech decoder which should receive speech frames at a fixed frequency. Note that the perceived effect of many recovery schemes and de-jittering buffer dynamics has been studied in literature [13, 14].

The PESQ algorithm has been basically designed to evaluate speech quality over telecom networks. In such a circumstance, the deletion of large speech sections (> 80 ms) is seldom observed. As such, PESQ algorithm will produce chaotic scores for degraded speech sequences subject to large loss instances. However, PESQ is sufficiently accurate to assess bursty sparse packet loss patterns and distorted speech sequences subject to loss instances with duration less than 80 ms [15]. Armed with this knowledge, our measurement space has been limited to MBLS and PLR values, respectively, equal to 80 ms and 30% (see Table 1). Moreover, we ensure that every loss instance is small than 80 ms. To fairly cover the whole packet loss space, the prevailing PLR and MBLS values of a generated packet loss pattern are checked. As a result, a synthesized trace is solely retained and considered when the deviation between specified and actual PLR and MBLS values are smaller than a given threshold.

**Table 1.** Empirical conditions for packet loss behavior using Gilbert model.

| Parameters | Conditions | Instances |
|---|---|---|
| CODEC | G.729 | 1 |
| Packet Loss Ratio (PLR) | 3, 5, 10, 12, 15, 20, 25, 30% | 8 |
| Mean Burst Loss Size (MBLS) | 1, 2, 3, 4 | 4 |
| Speech sequences | 16 male, 16 female | 32 |
| Total number of combinations | 1 × 8 × 4 × 32 | 1024 |

The measurement process is conducted using speech material that includes 32 standard 8 s-speech sequences, spoken by 16 male and 16 female English speakers. Such duration induces a maximal number of created 20 ms-voice packets equal to 400. Typically, such cardinality is insufficient to produce packet loss patterns with PLR and MBLS values close to theoretical values of PLR and MBLS set by users (see 'Appendix' section for further details). Moreover, unsent silence parts of a given speech sequence alter the initially generated packet loss pattern. This explains why we calculate and store the actual PLR and MBLS values for each couple of packet loss pattern and speech sequence (similarly as what it is done in [16] for video quality assessment). Table 1summarizes conducted experiences, where a total number of 1024 scores have been produced. As indicated in Table 1, we evaluate the performance of each SQA algorithm using the ITU-T G.729 coding scheme that is the unique speech CODEC covered by all examined speech quality assessors. It worth to note that our primary concerns is to examine the behavior and performance of bursty aware speech quality assessors under common configurations. In the scope of this work, the performance evaluation and improvement of speech CODECs under bursty packet loss processes are secondary concerns. A personalized extension of considered speech quality assessors to cover a large set of shared speech CODECs will be investigated in our future work using subjective tests.

## 4. SPEECH MATERIAL PREPARATION AND CONFIGURATION PARAMETERS SELECTION

A preparatory processing stage of speech material is necessary for a faithful assessment of speech quality. Indeed, manipulated raw speech sequence must meet a set of prerequisites for a consistent use of the ITU-T G.729 speech CODEC and the SQA algorithm defined in ITU-T Rec. P.862. In our case, raw speech material used to conduct our experiences was taken from the ITU-T P.Sup23 coded speech database [17]. The original sampling rate of considered speech sequences is equal to 16 kHz, where each sample is encoded using 16 bits. However, the specification of ITU-T G.729 speech CODEC indicated that input speech signals should be coded following linear PCM format characterized by a sampling rate and sample precision, respectively, equal to 8 kHz and 16 bits. As such, a down-sampling algorithm should be executed before processing speech signals by ITU-T G.729 speech CODEC. To do that, we resort to the open source and widely used software Sox (SOund eXchange) that comprises three distinguished resampling technology, a.k.a. frequency bandwidth changeovers, denoted as polyphase, resample, and rabbit strategies.

A dedicated SQA framework for the selection of suitable resampling technology has been set-up (see Figure 8). As we can observe, speech scores are artificially obtained using the full-reference ITU-T PESQ algorithm that can solely operate on speech signals sampled at 8 or 16 kHz. Note that the original and distorted speech sequences should be sampled at an equal frequency, i.e., either 8 or 16 kHz. Actually, the ITU-T PESQ algorithm is unable to score degraded speech sequences that incorporate fragments sampled at an unequal frequency. That is why each down-sampling operation should be followed by an up-sampling one. The features of considered speech material urge using the WB-PESQ algorithm that has been conceived for the evaluation of wideband coding schemes.

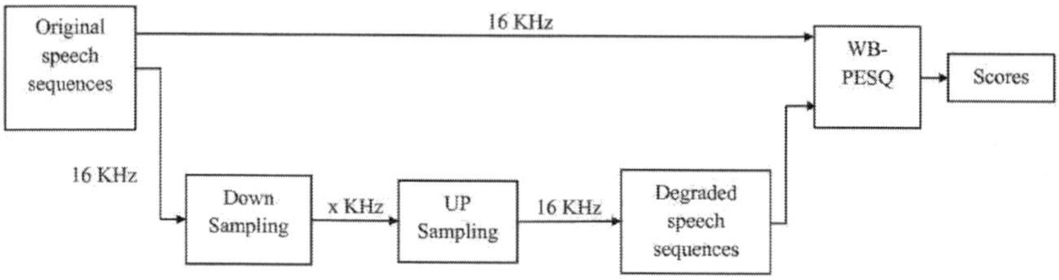


**Figure 8.** Framework for the evaluation of re-sampling technologies.

In Figure 8, we see that there is a possibility to evaluate multiple down- and up-sampling iterations using distinguished resampling technologies. Moreover, speech sequences are not coded to filter-out the effect of coding/decoding schemes. Actually, additional factors can interfere with resampling technology, such as filtering schemes, echo cancellers, de-noising algorithms, encoding schemes, and voice activity detectors. Moreover, configuration parameters of each re-sampling technology, such as window features, number of samples, and cutoff frequency influence its behavior.

A statistical analysis is applied to extract the perceived effect of resampling technologies. Figure 9gives some illustrative results about the perceived effect caused by the resampling technology using our set-up speech quality framework. Note that ITU-T WB-PESQ provides as a score a static value equal to 4.46 on MOS scale, when the two input speech signals are identical. Figure 9a illustrates the effect of one-iteration of up- and down-sampling iterations using polyphase and resample technologies on the treated speech sequences. As we can see, sampling technologies have distinct perceived effects following the speech content. The quality-degradation caused by the resampling technology is higher than the polyphase one. The average deviation of MOS-$LQO_{WB}$ between Poly-phase and Resample is equal to 0.1. As we can note, the quality-degradation is less perceptible for female sequences that are characterized by a high frequency. As a rule of thumb, the higher the final score, the smaller the quality deviation observed between examined resampling technologies. It seems that resampling technologies are less disturbing for speech waves characterized by a high frequency. Further tests indicate that the MOS-$LQO_{WB}$ scores are insensitive to the number of up- and down-iterations in a noiseless environment. Such an observation suggests that treated resampling technologies are roughly idempotent. In other words, the quality-degradation happens by resampling the original speech signals is null for already resampled speech signals.

The histograms given in Figure 9b present the average MOS-$LQO_{WB}$ scores produced by each treated re-sampling technology. As we can note, polyphase outperforms candidates resampling technologies. This explains why the polyphase resampling technology has been used to down-sample our original speech material.

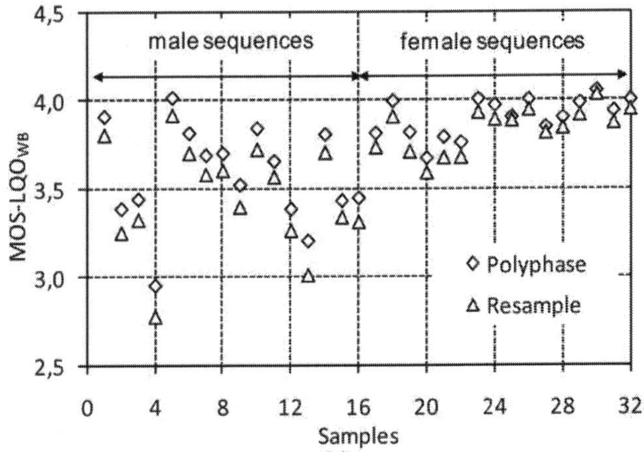


(a)

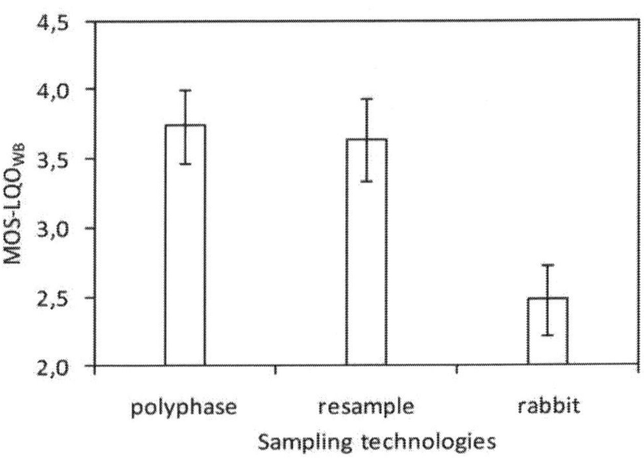


(b)

**Figure 9. Effect of re-sampling technologies on perceived quality**. **(a)** Effect of a 1-iteration of UP and DOWN sampling technology on MOS-LQOWB. **(b)** Average performance of sampling technologies as a function of MOS-LQOWB.

Apart the perceived effect of resampling technology, it is necessary to consider the VAD (Voice Activity Detector) algorithm included in ITU-T G.729 CODEC[h] to discriminate between active and silence speech wave sections [18]. This allows holding packet delivery processes during silence periods, which is highly recommended for the sake of utilization efficiency of network resources. The shortcoming of such a procedure consists of generating a mute-like signal between successive active periods in a way that could embarrass talker party. To generate more human-relaxing silence, ITU-T G.729 speech CODEC has been equipped with a CN capability. This option enables to periodically send at low rate Silence Insertion Descriptor (SID) packets that contain description about the ambient noise surrounding the listener party. As a result, the receiver will be able to generate more human-relaxing background noise.

For the sake of better quantification of perceived effect of CN mechanism, we conducted a preliminary series of experiences where eight reference speech sequences are distorted using a packet loss pattern generated following a Bernoulli distribution under activated and deactivated CN functionality. The average MOS-LQO scores of degraded speech sequences under enabled and disabled SID option are calculated for each loss condition. Under enabled SID option, loss instances that drop SID packets are ignored to emphasize their perceptual effect. The obtained results are plotted in Figure 10. As we can see, the overall LQ is basically insensitive to CN mechanism. In fact, considered speech sequences are gathered in a noiseless environment. This results in a little effect of CN mechanism on listening perceived quality. In reality, the CN mechanism should be explored in the context of considerable and time-varying background noises. This would allow developing smarter CN mechanisms that could be enabled/disabled according to prevailing background noises and packet loss processes. This will be considered in further detail in our future work.

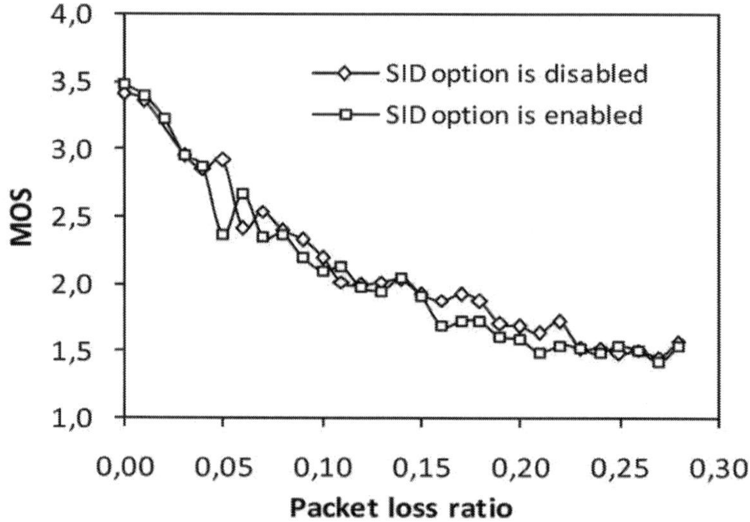


**Figure 10**. Effect of SID activation/deactivation on perceived quality under independent packet losses.

# 5. PERFORMANCE ANALYSIS OF BURSTY-LOSS AWARE SQA ALGORITHMS

In next sections, we start by describing calibrated parametric speech quality models that will subsequently enable an unbiased evaluation analysis. Next, we define our judgment metrics and discuss our findings. Notice that we assign the default values for various constants utilized by each speech quality assessor. To reach unbiased and consistent findings, the score yield by the explored SQA algorithms should be properly calibrated to satisfy the rating assumptions of PESQ algorithm. In fact, the designers of the PESQ algorithm calibrate its output to lay between that 1.5 to 4.5. That is why, we utilize existing quality models that has been derived using PESQ, rather than earlier subjective results [8, 19]. Precisely, for the VQmon and Q-Model assessment tools, we use the quality model given in (5) to estimate distortions due to independent packet losses. This model that is dedicated to the ITU-T G.729 speech CODEC has been obtained following a logarithmic regression analysis of PESQ scores under a wide range of PLR conditions [19]. The equation is

$$I_e = 22.45 + 21.14 \times \ln(1 + 12.73 \times \text{PLR}) . \tag{5}$$

As we can see from (5), under no loss condition, the utilized $I_e$ model induces a distortion amount equal to 22.45 rather than 11, which has been suggested based on earlier subjective-based testing [8]. Moreover, following ITU-T Rec. G.107, the values of $I_e$ should lay in the interval [0...40]. However, the $I_e$ model given in (5) can generate distortion measures as high as 73 for a PLR greater than 30%. Following our preliminary tests, this value may be considered as the upper bound that can be accurately obtained using PESQ algorithm. As such, for PLR values higher than 30% a value equal to 73 is assigned to $I_e$. For a fair comparison, we set, respectively, the lower and upper bound of the E-Model to 22.45 (no loss condition) and 73 (PLR higher than 30%). Further calibration is needless for Genome since it has been initially developed based on PESQ.

The metrics used to judge the performance of examined SQA algorithms are Pearson correlation coefficient and root mean squared error (RMSE) between measured and estimated rating factors, denoted hereafter respectively as ρ and Δ. The value of Δ is obtained using the following expression:

$$\Delta = \sqrt{\frac{1}{N}\sum_{i=1}^{N}\left(R_{\mathrm{M}}^{i} - R_{\mathrm{E}}^{i}\right)^2}, \tag{6}$$

where, $R_{\mathrm{M}}$ and $R_{\mathrm{E}}$ refer, respectively, to measured and estimated rating factors and $N$ is the number of measures. The conducted measurement study evaluates rating performance according to the following two perspectives:

- *Sequence-by-sequence methodology*: It consists of directly computing ρ and Δ values using the measured and correspondent estimated scores. This strategy enables some understanding of the sensitivity of a given SQA algorithm with respect to a specific bursty packet loss pattern and the speech content of a given sequence.

- *Cluster-by-cluster methodology*: It consists in creating a set of groups of measured scores according to shared features, such as PLR, MBLS, active and silence durations. For each measure and examined SQA algorithm, the estimated score is inserted into the corresponding group of the measured cluster. Finally, we calculate the average of measured and estimated scores of each produced cluster. The values of ρ and Δ are obtained by processing averaged scores of clusters. This strategy enables to filter-out deviations caused by speech content and specific packet loss distributions that may be required to satisfy specific needs of some applications and service providers, especially for planning purposes.

In the following, E-Model (1) and E-Model(2) denote, respectively, the E-Model designed to consider independently and bursty dropped packets [3]. Q-Model (1) and Q-Model (2) refer, respectively, to the Q-Model where local burstiness increases linearly and exponentially, as a function of inter-loss gap (see 'Genome' section) [11].

Histograms given in Figure 11a summarize the obtained value of ρ using sequence-by-sequence and cluster-by-cluster measurement strategies. Each cluster comprises scores obtained for a given measured PLR range independently of the MBLS values and speech contents. The width range of PLR values covered by each cluster is equal to 5%. As we can see in Figure 11a, all SQA nearly achieve a perfect correlation coefficient under cluster-by-cluster measurement strategy. The correlation coefficients are slightly inferior using a sequence-by-sequence measurement strategy. This observation is somehow expected, as a significant increase of PLR values induces a considerable decrease of MOS scores, and conversely. All existing SQA algorithms are designed using monotonic quality models as functions of PLR values, which explains the observed good correlation coefficients. This feature is more emphasized for the cluster-by-cluster measurement methodology, since it eliminates unusual deviations caused by a specific bursty packet loss pattern and speech content. As we can see, Q-Model (1) and Q-Model (2) slightly outperform other SQA approaches. Moreover, we see that VQmon achieves the minimum correlation coefficient following our measurements.

Histograms given in Figure 11b summarize the obtained values of Δ using sequence-by-sequence and cluster-by-cluster measurement strategies. As we can see, the examined SQA algorithms induce significant deviation between measured and estimated scores. E-Model(1) induces the maximal value of mean deviation, which is expected since it has been designed for randomly removed packets. Q-Model(2) achieves the minimum average deviation. The accuracy of E-Model(2) is better than E-Model(1)'s since it subsumes more properly packet loss burstiness. As we can note, the minimum value of Δ is roughly equal to 6, which in our opinion is still pretty important. This constitutes the principal weakness and limitation of the treated SQA, which should be comprehensively tackled in future work.

For a deeper understanding of the behavior of the examined four SQA algorithms, in Figure 12 we provide scatter plots that visually illustrate the correlation and accuracy of estimated scores. As we can see, Q-Model(1) and Q-Model(2) exhibit superior behavior rating than other SQA algorithms (see '◊' symbols located more closely to the $y = x$ line). Moreover, we note the presence of certain outliers that significantly deviate from measured scores, which are more significant for VQmon. Furthermore, we can see that E-Model(1), Genome, VQmon, and Q-Model(1) tend to overestimate the measured scores. However, the trend of E-Model(2) is to over- (resp. under-) estimate measured scores under small (resp. high) PLR values. This signifies that an additional calibration process can surely improve the output accuracy of SQA algorithms. For the sake of explanation, a first-order linear regression process has been applied on the obtained raw dataset. Table 2 illustrates that the calibration process notably improves the estimation accuracy (< 6) while keeping exactly the same correlation coefficient. The transformed score of the *i* th measure is given by:

$$R_{\mathrm{T}}^{i} = aR_{\mathrm{R}}^{i} + b, \tag{7}$$

where *a* and *b* are the fitting coefficients that minimize the RMSE. $R_{\mathrm{T}}$ and $R_{\mathrm{R}}$ stand for transformed and raw rating factors, respectively. As we can see, Q-Model(1) and Q-Model(2) slightly outperform other competing strategies. The transformed (improved) models can be utilized for a better estimation of measured rating factor.

The performance metrics previously calculated consider all measurements at once, which may lead to ignore/hide some specific features of the examined SQA algorithms. For the sake of enlightenment, we calculate the values of ρ and Δ using striped dataset scores following the value of PLR. Precisely, each dataset strip comprises scores that have been observed for a PLR range equal to 10%. Figure 13 illustrates the values of ρ and Δ for each dataset strip. As we can see, bursty-aware SQA algorithms exhibit an acceptable correlation under small (< 10%) and high (> 20%) packet loss ratios. However, there is a clear trouble to estimate scores under moderated PLR values (10 to 20%). From Figure 13b, we see that the values of Δ are quite large under all conditions. Moreover, as expected, E-Model(1) achieves an acceptable correlation under light loss process, where voice packets are independently/sparsely deleted. However, the efficiency of E-Model(1) sharply decreases as packet loss severity increases. E-Model (2), Q-Model(1), and Q-Model(2), which are the bursty aware varieties of E-Model(1), provide more accurate and correlated scores. These results revealed that Q-Model(2) achieves best trade-off between correlation and accuracy.

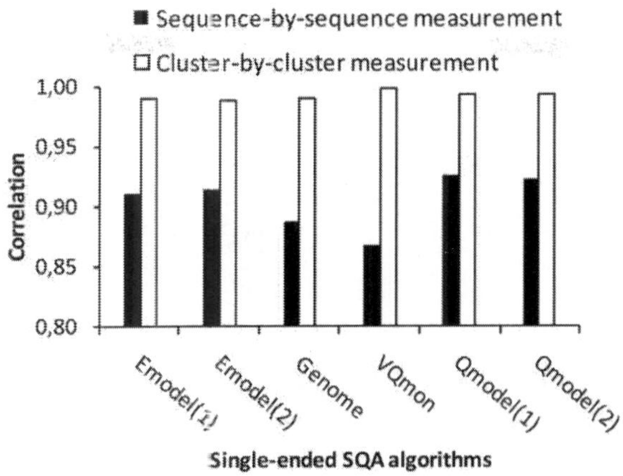


(a)

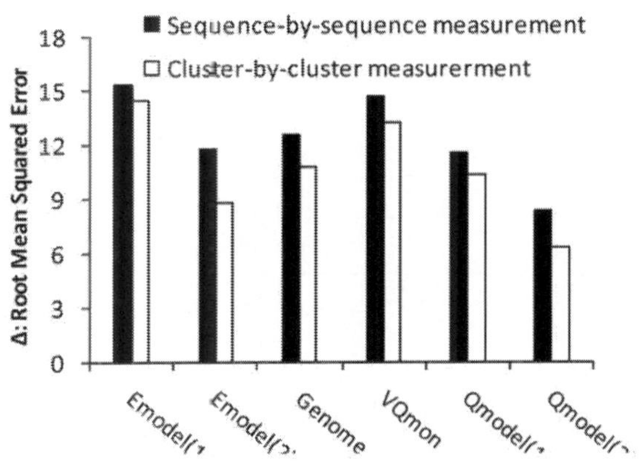


(b)

**Figure 11. Correlation factor and average deviation on sequence-by-sequence and cluster-by-cluster bases under limited bursty packet loss space**. **(a)** Correlation between measured and estimated measures. **(b)** Mean deviation between measured and estimated.

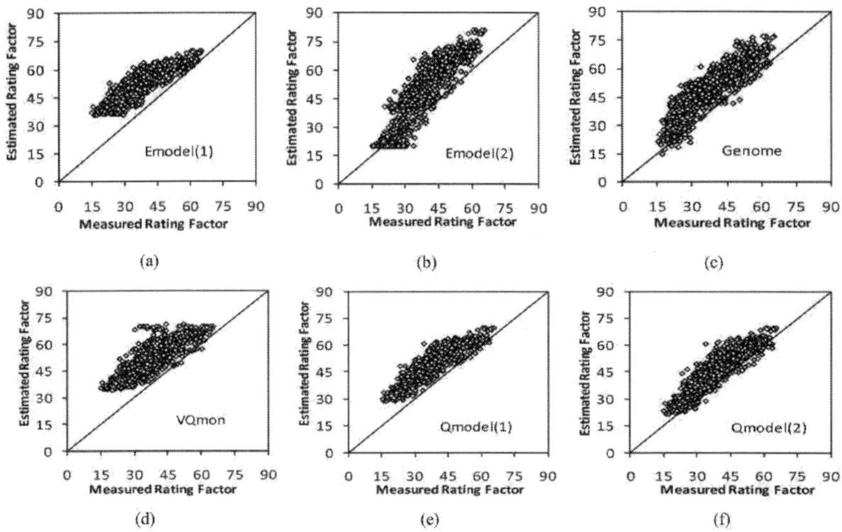


**Figure 12.** Relationship between measured and estimated scores through scatter **plots. (a)** Emodel(1), **(b)**Emodel(2), **(c)** Genome, **(d)** VQmon, **(e)** Qmodel(1), and **(f)** Qmodel(2).

**Table 2.** Summary of calibrated models and their performance.

| SQA algorithm | *a* | *b* | ρ | Δ |
|---|---|---|---|---|
| E-Model(1) | 1.170 | -23.016 | 0.91 | 4.738 |
| E-Model(2) | 0.607 | 9.066 | 0.91 | 4.664 |
| Genome | 0.821 | -2.740 | 0.89 | 5.324 |
| VQmon | 0.965 | -11.694 | 0.87 | 5.741 |
| Q-Model(1) | 1.017 | -11.466 | 0.92 | 4.380 |
| Q-Model(2) | 0.872 | -1.344 | 0.92 | 4.473 |

Besides the limited previously explored space, we conducted with precaution some experiences in order to evaluate the performance of bursty-aware SQA algorithms over a wide range of conditions. The values of PLR (resp. MBLS) have been varied from 5% (resp. 1

packet) to 40% (resp. 10 packets). A total number of combinations equal to 2240 have been evaluated. Table 3 summarizes the obtained values of ρ and Δ on sequence-by-sequence basis. The pertinent observed feature is the high value of Δ. This is somehow expected since neither the full-reference SQA algorithm ITU-T Rec. P.862 nor examined bursty-aware SQA are designed to evaluate loss conditions characterized by large losses instances (> 80). In [20], a proposal for a novel speech quality assessor has been introduced that considers more properly this problem.

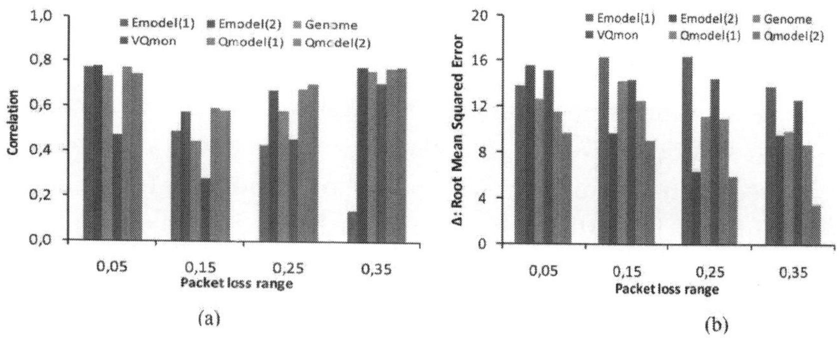


**Figure 13. Performance judgment metrics as function of PLR range under a limited bursty packet loss space. (a)** Correlation on interval basis. **(b)** Deviation on interval basis.

**Table 3.** Performance of bursty-aware SQA algorithms under a large space.

| SQA algorithm | ρ | Δ |
|---|---|---|
| E-Model(1) | 0.940 | 14.273 |
| E-Model(2) | 0.898 | 18.488 |
| Genome | 0.882 | 15.465 |
| VQmon | 0.913 | 14.634 |
| Q-Model(1) | 0.938 | 14.338 |
| Q-Model(2) | 0.929 | 15.125 |

# 6. CONCLUDING REMARKS AND PERSPECTIVES

The learned lessons of our performance analysis of bursty-aware SQA algorithms can be resumed as follows:

1) Existing bursty-aware SQA algorithms are basically designed to averagely approximate the subjective score of a given disturbing configuration. This signifies that they are unsuitable to accurately estimate speech quality on a sequence-by-sequence basis.

2) The strategy of the Q-Model achieves a consistent and reasonable performance under a wide range of conditions. Further investigation is necessary for a better and dynamic calibration. The Q-Model assures an elegant trade-off to subsume the perceived effect of packet loss at short- and long-terms. In our opinion, it constitutes a solid base for the development of a sequence-by-sequence SQA strategy, which considers speech content, packet loss burstiness, and 'recent' effect.

3) VQmon and E-Model(2) need more improvement to accurately judge perceived quality. Indeed, they seem to be more suitable for assessments over long periods since they utilize characterization parameters that need an important amount of measures to be stabilized. Moreover, both strategies definitely ignore temporal distribution details of loss instances.

4) The statistical property of Genome leads to some inaccuracy in the estimated scores. Preliminary conducted experiences revealed that it is insensitive to the distribution of (inter-loss, loss) couples.

As future work, we strongly believe that a hybrid speech quality assessor that utilizes additional meta-data about speech wave are required to improve accuracy of existing SQA algorithms such as silence/active patterns and feature of removed signals, e.g., voiced or unvoiced. Moreover, the location of a given loss instance should be considered during the evaluation processes. We believe that a perceptual packet loss pattern should be determined according to the concrete packet loss pattern and sequence features. Furthermore, it is crucial to extend existing speech quality assessors to cover a wide range of speech CODECs using subjective tests under longer bursty packet loss processes. This will enable identifying which assessment methodology is better as a function of the running speech coding scheme. The goal is the development of a versatile and highly accurate speech quality assessor of VoIP service on call-by-call basis.

Finally, it is important to note that the authors realize that extensive subjective testing should be done to tune, validate, and improve the competitive speech-quality assessment technologies. This constitutes a principal priority that will be addressed in our future work.

# APPENDIX

On Packet Loss Modeling over VoIP Networks

The metrologies of packet loss throughout VoIP calls show that voice packets are removed in bursts. Basically, bursty packet loss processes are modeled using either discrete- or continuous-time Markov chains. A simple, yet accurate 2-state discrete-time Markov chain, referred to as Gilbert model, or sometimes simplified Gilbert model, has been well explored in the literature (see Figure S1a, Additional file 1) [21]. It was proposed to analyze noisy channels that introduce bursty bit errors. It has been subsequently extended to model bursty packet loss processes [21].

In a few words, Gilbert model has NO-LOSS and LOSS states that respectively represent successful and failing packet delivery operations. The Gilbert model is fully characterized by its transition probabilities $p$ and $q$ (see Figure S1a, Additional file 1). For sake of clarity, the model is instead characterized using Packet Loss Ratio

(PLR) and Mean Burst Loss Size (MBLS). The following relationships enable the mapping between characterization parameters:

$$p = \frac{\text{PLR}}{\text{MBLS} \times (1 - \text{PLR})} \quad \text{and} \quad q = 1 - \frac{1}{\text{MBLS}}. \tag{8}$$

Besides capturing the features of bursty packet loss processes, the Gilbert chain can be utilized to synthesize packet loss patterns following user-defined PLR and MBLS values. Notice that a large number of packets should be generated to produce packet loss patterns that respect PLR and MBLS values given by the user. Figure S2, Additional file 1 illustrates the average deviation between specified and measured PLR and MBLS of ten generated packet loss patterns using distinct seed values, as a function of the number of generated packets. As we can observe, the greater the number of generated packets, the lower the deviation between specified and measured PLR and MBLS. This series of experiences showed that number of packets greater than 3000 packets achieves sufficient accuracy between target and measured PLR and MBLS values.

Besides this discrete-time Gilbert model, a continuous-time 2-state Modulated Markov Poison Processes (MMPP-2) can be used to characterize time-varying packet loss processes that alternate between low and high packet loss periods (see Figure S1b, Additional file 1). In state 0 (resp. 1), packet loss instances are introduced to the rendered packet stream following Bernoulli processes with average value equal to $\text{PLR}_{\text{LOW}}$ (resp. $\text{PLR}_{\text{HIGH}}$). The parameters of the MMPP-2 model can be estimated at run time for a given data trace using a maximal likelihood estimator (MLE) [22]. Multiple variants of the expectation-maximization (EM) algorithm have been utilized by statisticians to obtain such values [23]. Li [23] developed a freely downloadable code of a variety of EM algorithms dedicated to calibrate MMPP model. The calibrated model can be utilized to judge the severity of packet loss burstiness and its variability.

To generate packet loss patterns using the MMPP-2 model, the PLR values can be randomly selected at the start time of each new period among a set of user-defined values. The sojourn period in each state follows an exponential distribution that should be parameterized by users. Figure S3, Additional file 1 shows multiple profiles generated using the MMPP-2 model described previously under several settings. As we can observe, MMPP-2 produces more realistic packet loss profiles under a large observation interval.

The previously described Gilbert and MMPP models give coarse features of time-varying and bursty packet loss process. As such, packet loss patterns that could lead to misestimating the perceived quality are poorly considered. To enable a better characterization, Clark [5] proposed a dedicated packet loss model that discerns between loss instances happen in gap and in burst (see Figure S4, Additional file 1). As we can see, Clark's model has four states labeled 1, 2, 3, and 4. The sub-chain 1 is used to consider isolated packet loss instances. However, the sub-chain 2 is used to consider temporally dependent packet loss instances. The author defines the following two triggering conditions to switch from sub-chain 1 to sub-chain 2:

1) A loss instance that comprises more than two consecutive missing packets.

2) A single missing packet preceded by a loss event that has been happened at a distance smaller than a given constant $g_{min}$. Clark recommends using a value equal to 16 10-ms voice packets.

A transition from sub-chain 2 to sub-chain 1 happens once an isolated packet loss instance preceded by $g_{min}$ successfully received packets is detected. Clark [5] developed an efficient packet loss driven algorithm that enables to calibrate at run-time the proposed model. A set of metrics can be extracted from Clark model at the end of a monitoring period, e.g., PLR during gap and bursty loss periods and their corresponding durations. As depicted in Figure S4,

Additional file 1, Clark accounted for the effect of discarded packets at the de-jittering buffer caused by late arrivals.

## ENDNOTES

a) A loss instance is defined as a block of consecutive missing packets delimited by two successfully received ones.

b) The initial version of VQmon suggests the use of time constants $\tau_1$ and $\tau_2$, respectively, equal to 5 and 15 s [4]. Recently, a more elaborated analysis conducted by Raake [3] indicated that time constants $\tau_1$ and $\tau_2$, respectively, equal to 9 and 22 s are more accurate to mimic users' behavior rating.

c) This definition implies that the delivery network introduces independent (resp. bursty) packet losses when BurstR is equal to (resp. greater) one. As a rule of thumb, the greater the value of BurstR above 1, the higher the intensity of packet loss burstiness. Notice that MBLS value of the expected independent packet loss processes is equal to 1/(1 - PLR) where the value of PLR is set to the measured packet loss ratio.

d) The variable CLP refers to the probability of losing a packet given that the previous one is lost.

e) A packet loss process that periodically drops a static number of consecutive speech frames preceded by a given inter-loss gap size.

f) Precisely, the value of $\alpha_n$ is set to 1 if packet loss ratio till *n* th packet is below 4%, otherwise it is set to -1/2.

g) The recommended value of window size is equal to 8 20-ms voice packets.

h) Basically, all emerging speech CODEC include a built-in VAD.

## ACKNOWLEDGEMENTS

We would like to express our sincere thankfulness to the anonymous reviewers for their constrictive comments that helped us to improve the paper during the submission processes. In particular, the authors feel committed to pursue investigation in some specific issues according to reviewers' recommendations.

## REFERENCES

1. Rix A, Beerends J, Kim D, Kroon P, Ghitza O: Objective Assessment of Speech and Audio Quality: Technology and Applications. *IEEE Trans Audio Speech Language Process* 2006, 14(6):1890-1901.

2. Jelassi S, Youssef H, Pujolle G: Perceptual Quality Assessment of Packet-Based Voice Conversations over Wireless Networks: Methodologies and Applications. In *Quality of Service Architectures for Wireless Networks: Performance Metrics and Management*. IGI Global Publisher; 2009.

3. Raake A: Short- and long-term packet loss behavior: towards speech quality prediction for arbitrary loss distributions. *IEEE Trans Audio Speech Language Process* 2006, 14(6):1957-1968.

4. Mohamed S, Rubino G, Varela M: Performance evaluation of real-time speech through a packet network: a random neural networks-based approach. *Perform Eval* 2004, 57(2):141-162.

5. Clark A: Modeling the effects of burst packet loss and recency on subjective voice quality. In *Proceedings of 2nd IP-Telephony Workshop (IPTel'2001)*. Columbia University, New York City, USA; 2001.

6. ITU-T: Study the relationship between instantaneous and overall subjective speech quality for time-varying speech sequence: influence of a recency effect. 2000.

7. Jelassi S, Youssef H, Hoene C, Pujolle G: Voicing-aware parametric speech quality models over VoIP networks. In *Proceedings of 2nd IEEE Global Information Infrastructure Symposium (GIIS 2009)*. Hammamet, Tunisia; 2009.

8. ITU-T: The E-Model, a computational model for use in transmission planning. *Recommendation G.1072*005.

9. Cole RG: JH Rosenbluth Voice over IP performance monitoring. In *Comput Commun Rev. Volume 31*. ACM SIGCOMM; 2001:9-24. 10.1145/505666.505669

10. Roychoudhuri L, Al-Shaer E: Real-time audio quality evaluation for adaptive multimedia protocols. In*Proceedings of Multimedia Networks and Services (MMNS 2005)*. Spain; 2005.

11. Zhang H, Xie L, Byun J, Flynn P, Shim C: Packet loss burstiness and enhancement to the E-Model. In*Proceedings of the 6th IEEE International Conference on Software Engineering, Artificial Intelligence, Networking and Parallel/Distributed Computing*. Towson, Maryland, USA; 2005.

12. ITU-T: Perceptual evaluation of speech quality (PESQ): an objective method for end-to-end speech quality assessment of narrow-band telephone networks and speech codecs. *Recommendation P.862* 2001.

13. Turunen J, Loula P, Lipping T: Assessment of objective voice quality over best-effort networks. *Comput Netw* 2005, 28(5):582-588.

14. Jelassi S, Youssef H, Pujolle G: Parametric speech quality models for measuring the perceptual effect of network delay jitter. In

*Proceedings of 34th Annual IEEE Conference on Local Computer Networks (LCN 2009)*. Zürich, Switzerland; 2009.

15. Basterrech S, Rubino G, Varela M: Single-sided real-time PESQ score estimation. In *Proceedings of Measurement of Speech, Audio, and Video Quality in Networks (MESAQIN2009)*. Prague, Czech Republic; 2009.

16. Couto-da-Silva A, Rodriguez-Bocca P, Rubino G: Optimal quality-of-experience design for a P2P multi-source video streaming. In *Proceedings of ICC'08*. Beijing, China; 2008.

17. ITU-T: Coded-speech database. *Recommendation P.Supplement 23* 1998.

18. ITU-T: Coding of speech at 8 kbit/s using conjugate-structure algebraic-code-excited linear prediction (CS-ACELP). *Recommendation G.729* 2007.

19. Sun L, Ifeachor E: New models for perceived voice quality prediction new models for perceived voice quality prediction optimization for VoIP networks. In *Proceedings of IEEE International Conference on Communications (ICC 2004)*. Paris, France; 2004:1478-1483.

20. Jelassi S, Youssef H, Sun L, Pujolle G, NIDA: a parametric vocal quality assessment algorithm over transient connections. In *Proceedings of 12th IFIP/IEEE International Conference on Management of Multimedia and Mobile Networks and Services (MMNS 2009)*. Venice, Italy; 2009.

21. Sanneck H: *Packet Loss Recovery and Control for Voice Transmission over the Internet*. Technical University of Berlin; 2000.

22. Rydenl T: An EM algorithm for estimation in Markov-modulated Poisson processes. *Elsevier Comput Stat Data Anal* 1992, 21(4):431-447.

23. Hui L: Workload Modeling in Grid Computing Environments. 2010.

24. Carvalho L, Mota E, Aguiar R, Lima AF, de Souza JN, Barreto A: An E-Model implementation for speech quality evaluation in VoIP systems. In *Proceedings of the 10th IEEE Symposium on Computers and Communications (ISCC'05)*. La Manga del Mar Menor, Cartagena, Spain; 2005.

3

# A TOPOLOGY-AWARE RELAY LOOKUP SCHEME FOR P2P VOIP SYSTEM

**Xiuwu ZHANG[1], Ruifeng GUO[1,2], Weimin LEI[1,2], Wei ZHANG[1]**

[1]School of Computer Science and Technology, University of Science and Technology of China, Hefei, China

[2]Shenyang Institute of Computer Technology, Shenyang, China

## ABSTRACT

Because of the best-effort service in Internet, direct routing path of Internet may not always meet the VoIP quality requirements. Thus, many researches proposed Peer-to-Peer VoIP systems such as SIP+P2P system, which uses relay node to relay RTP stream from the source node to the destination node and uses application-layer routing scheme to lookup the best relay nodes. The key of those systems is how to lookup the appropriate relay nodes, which we call relay lookup problem. This paper presents a novel peer relay lookup scheme based on SIP+P2P system. The main ideas are to organize the P2P network using a Cluster overlay and to use topology-aware to optimize relay selection. We introduce the mechanism in detail, and then evaluate this mechanism in NS2 network simulation environment. The results show that our scheme is scalable and can get high relay hit ratio, which confirm the feasibility of a real system. We also make comparison with traditional schemes and the results show that our scheme has good path quality.

## KEYWORDS

Relay Lookup, Cluster, QoS, Peer-to-Peer VoIP

## 1. INTRODUCTION

Unfortunately, the current Internet provides a Best-effort service, which cannot guarantee VoIP service qualities on the IP-layer sometimes. That is, sometimes direct routing path in the Internet may not always provide good path quality. For the good quality of service, it is important to select alternate paths between the source node and the destination node in the application layer. Thus, it has been popular to build overlay routing for VoIP application based on Peer-to-Peer (P2P) networks in the Internet [1–3].

Many previous research works [2,4,5] show that overlay routing can effectively improve VoIP session's performance by avoiding IP-layer's path failure. In application-layer routing scheme for VoIP systems, we should first lookup appropriate relay nodes in the application layer to bring an intermediate routing scheme. We call it the relay lookup problem. The definition of relay node is one which can not only give relay services for local users, but also serving as a bridge for remote nodes in the topology. The role of relay node is not only to support two/multiple node's communication (traverse a NAT or firewall), but also to avoid congestion or failure in IP-layer to improve the quality of communication capacity. Many researches [2,6] show that relaying RTP packet is an effective way to improve the quality of VoIP communication.

There are many application-layer routing projects which propose relay lookup scheme, such as MIT's resilient overlay network (RON) [4,5], Ohio state University's AS-Aware Peer-Relay Protocol (ASAP) [1], which all are to solve the Internet path failure through one-hop overlay routing. Besides these systems, there is one commercial software which is Skype [3,7,8]. As we know, Skype successfully use

P2P technology and supports millions of online users, and the most important technique is the relay mechanism. Skype profits from the relay mechanism, but the fact is that, Skype encrypts its packets so that the exact routing methods are unknown. These defects hinder its further growth.

In our initial work, SIP+P2P system [2], we propose a P2P relay mechanism, which is to make those nodes with good network situation and machine performance become relay nodes, and to use a P2P overlay to organize those nodes. In the original system, we use a simple relay lookup scheme which we call one-side optimal selection scheme. In one-side optimal selection, client nodes cache a list of relay nodes and keep heartbeat detecting. When need relays, the caller client node selects several nearest relay nodes in cache by means of RTT and negotiates with peer client to find proper relays. But this mechanism is a unilateral optimal method, not suitable for bilateral optimal requirement in VoIP communication.

Thus, how to design an effective scheme in SIP+P2P system to find the appropriate nodes is really a field worth researching. In this paper, we propose a novel peer relay lookup scheme based on cluster participation, which can be used in SIP+P2P system. The main idea is to use topology-aware to optimize the peer relay selection. We divide the whole P2P relay network into many clusters, and organize the whole P2P network using Cluster overlay. When one node joins the system, it will automatically upgrade its identity to be a landmark node or just be an ordinary node according to its own network situation and relative location. If it is upgraded to be a landmark node, it will be in charge of building clusters in the overlay network, and other ordinary nodes will join in the cluster. The landmark node is the responsible one of the cluster, and it will record the relay nodes in its cluster which have the ability to relay with other clusters. When the node in the cluster needs relay nodes during its communication, the landmark will give out a relay node candidate list, and the communication node chooses the right relay node from the list after testing and negotiating with peer node. In the aspect of cluster topology constructing, we use path quality as the measurement metric, which is quite different from ASAP. ASAP use AS information from BGP routing table to build cluster topology, which is unrealistic in large scale network and mobile network. Instead, our measurement metric of path quality also reflects the proximity relationship in topology distance, and is easy to realize and suitable for large scale network.

## 2. RELATED WORK

The resilient overlay network (RON) [4,5] is a solution for general application-layer routing problems, which aims to provide optimized packet routing performance through routing packets in an overlay network. Using link state routing protocol, RON server detects all the links in regular cycle, thus it gets the update link state of the whole network, including the delay between any paired nodes. When there is a failure in the direct path, the nodes ask the server to find relay nodes. The server will check all the links around the source node and the destination node, and then return a proper node list which is near to both nodes. RON is only efficient for small network application in which the nodes number is below 50. Besides, RON is not specifically designed for VoIP system, thus, is not suitable for relay service in VoIP system.

Many researchers attempt to gain insights into the Skype relay selection algorithm in a black-box manner [3, 78]. They capture the packets to study the protocol in Skype and analyze the key techniques according to these traces. From the perspective of specific technologies, Skype's application layer routing has several advantages, including quick routing, lower operating costs and network overhead. Then the system can greatly reduce the burden of the central server. Nevertheless, on the one hand, the relay selections do not take topology into consideration, so it cannot always find proper relays. On the other hand, due to its private signal and encrypting packet transmission, we still do not know the detailed design of the system up to now.

Our previous work, SIP+P2P system [2], proposed a P2P relay mechanism to improve the VoIP quality. We select nodes which have good network situation and machine performance to become relays and provide relay service for VoIP sessions. Then we use a P2P overlay to organize those nodes. In our original system, we used a simple peer relay lookup scheme: all client nodes cache a list of relay nodes and keep alive; when a node needs relays to communicate, it will select several nearest relays and negotiate with the peer node; after detecting and negotiating, the two nodes find appropriate relay nodes for their communication. In order to optimize the selection procedure, client node evaluates relay nodes in its cache by means of ping/pong RTT value, and sorts them according to the evaluation result. We also classify all relay nodes into different ISP sets to enhance path diversity and reduce the cache space. For all that, this mechanism can-not always find the near relays for both sides yet, because it is a unilateral optimal selection. But VoIP systems often require that the selected relays are near to both the source node and the destination node. Thus this relay lookup mechanism needs to be further improved.

ASAP (AS-Aware Peer-Relay Protocol) [1] is specially designed for VoIP systems and as it experienced, they got a good result for VoIP communication nodes. The main idea is to divide the whole network into AS domains. According to IP prefix, the system classifies all nodes that have joined the system into different ASs, which are called clusters. There are Bootstrap servers in the system in charge of the IP prefix of clusters and the overall AS graph. There are responsible nodes, also called proxy nodes, in every cluster. These proxy nodes are chosen from the clusters with better network state. They get the location of AS and the IP prefix table from the Bootstrap server, and then make a record of nodes in the cluster. When nodes join the system, they can find their cluster nodes based on its own IP, and let the cluster record. Also proxy node will detect the surround AS and record the near ones. When two communicating nodes need relays, they will ask proxy nodes for the near clusters' information and the two clusters' information will make an intersection and the nodes in the set will be the relay candidates. The two nodes will then detect the relay candidates to find the appropriate relays. The key of this system is to find the IP prefix to give nodes basic judgment to join the right cluster. But getting the IP prefix from the BGP routing table is not easy to bring out for large scale network, and the real-time update one is even more difficult and needs lots of work, so the actual deployment of system in large scale network is unrealistic. But the system itself is great reference for relay lookup mechanism.

As introduced above, both Skype and SIP+P2P system have some limits: 1) their peer selection is suboptimal; 2) there are a large number of unnecessary probes. Those systems can be further improved by using topologyaware technology. ASAP uses AS-Aware to optimize relay selection, but it still has some limitations. Thus, we propose a novel peer relay lookup based on SIP+P2P system, which still uses a Peer-to-Peer relay mechanism. We organize the P2P relay network using Cluster overlay, and use topology-aware to optimize relay lookup. When build cluster overlay and select relay, we use path quality as its measurement metric, which is easier to realize in large scale network.

# 3. DESIGN OF PEER RELAY LOOKUP SCHEME

Our design still confirms to the framework of SIP+P2P system, in which the SIP signal is client/server based, but the media stream is transmitted in P2P relay network. We still adopt one-hop relay scheme, and use topologyaware technology to help selecting the proper relays. We divide the whole relay network into different clusters according to the path quality, which denotes the distance between two nodes. Thus, nodes which have good quality path (e.g. MOS>3.5) to each other are near in topology and will join in the same cluster. We use delay and packet loss to evaluate the path quality, because it is generally thought that the voice quality mainly relates to the delay and the packet loss among all network factors. When the one-way delay is lower than 200ms and the packet loss less than 2%, the voice quality is usually acceptable. In this cluster system, there are three kinds of nodes:

- Landmark node: Landmarks are chosen from ordinary nodes when they have good hardware condition and network surrounding. The landmark is the management center of the cluster.
- Relay node: Relay nodes provide relay service for other nodes. They usually have good network situation. When a node becomes a relay, it will record its address on landmark.
- Ordinary node: Ordinary nodes will join the cluster by being recorded on the landmark and maintain the states with periodic heartbeat method.
- Relay nodes and ordinary nodes are all cluster nodes. Sometimes, a landmark node may also be a relay node. When communication nodes need relaying, they will ask the recording landmark for proper relay nodes candidates, and then detect the relay to see if it can be the relay node.

## 3.1. System Architecture

As shown in **Figure 1** and **Figure 2**, the system is composed of some Bootstrap Servers, a cluster layer and a logical relay layer. The Bootstrap Server is only serverlike component in the system which needs to be deployed previously. It is responsible for node login, landmark management, and maintains a bootstrap's data structure, including: a cluster graph, a cluster landmark IP table. Cluster layer is used to organize nodes and maintain the real topology. Cluster layer is formed with several clusters. In each cluster, nodes are tagged as landmarks or cluster nodes. Each cluster node is attached to only one landmark. In normal cases, the quality of path between a node and its landmark is very good. The landmark node maintains its near clusters to form the topology, as show in **Figure 1**. Relay logical layer is used to help nodes in cluster to find appropriate relay nodes. Relay layer is a virtual layer with all nodes mapped from cluster layer; the connection relationship is kept the same as the one in cluster layer, simply adding some relay related information on each node to help find relay.

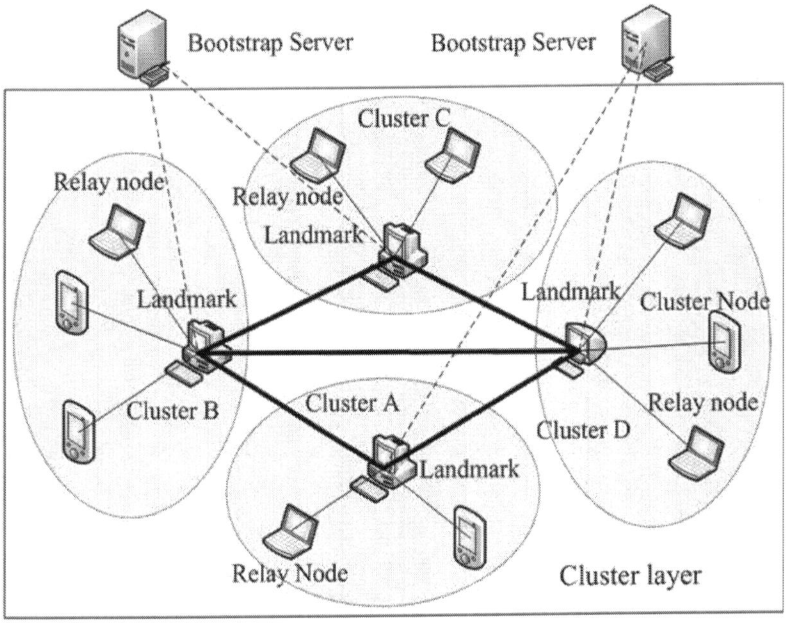


**Figure 1**. Deployment of system architecture.

## 3.2. Module Design on the Peer

**Figure 3** shows the module design on one peer. There are two main modules, including cluster management module and relay module. There are two parts in cluster management module: the landmark logic and the cluster node logic. And relay module also consists of the relay server logic and the relay client logic.

Cluster management module: Its main function is to maintain the cluster topology. The landmark logic maintains a neighbor landmark set, a cluster node set and a relay node set. Landmark logic is opened when this node becomes a landmark. The cluster node logic maintains its landmark set and a standby landmark set. There will be one candidate landmark which is decided by landmark competing. All nodes keep heartbeat communications with other nodes, and will help them to find the relay nodes candidates. The node will change the landmark set according to the dynamic network states.

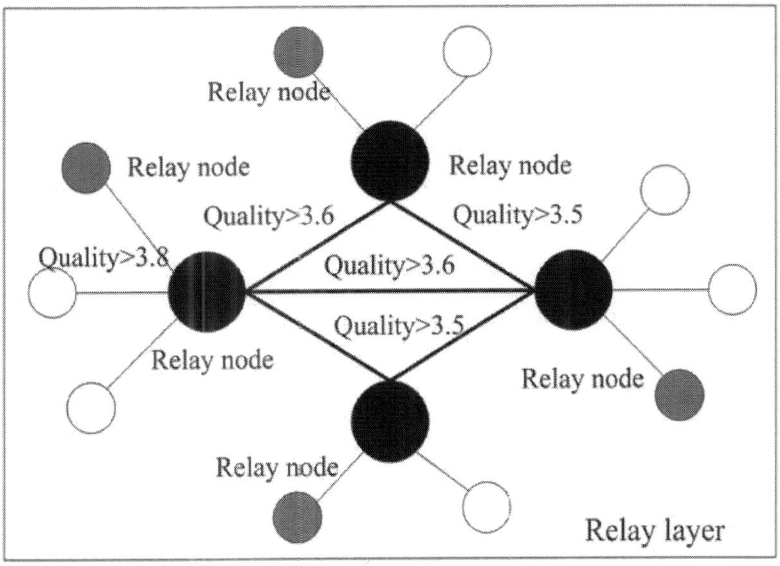


**Figure 2**. A relay logical layer mapped from the clusterlayer.

Relay module: The main function is to provide relay service for upper SIP sessions. Relay server logic is an independent function. When a cluster node can become a relay, it will open this function and record its address on the landmark. Relay client logic mainly helps to find candidate relay nodes by detection. At the same time, it will cache a local relay list which has been found to be the proper relays. But the relay client will first ask the cluster management module to find proper relay nodes; if it is failed, it will then use the local relays.

## 3.3. Node Operations

Node joining: There is a fixed address list of the bootstrap servers stored in every client. The cluster node will send a RTT detection request to the bootstrap server and receive the response when it logins. Then according to the RTT, the normal node requests a list of landmarks from the nearest bootstrap server. After got a list of landmarks, this node detects them and finds out the nearest landmark according to the path quality. We set a threshold when judging whether a node belongs to a cluster. If there is no corresponding cluster landmark, this node will be automatically promoted to be a landmark to form a new cluster, and then will get a cluster ID from the bootstrap server which is used to represent the new cluster.

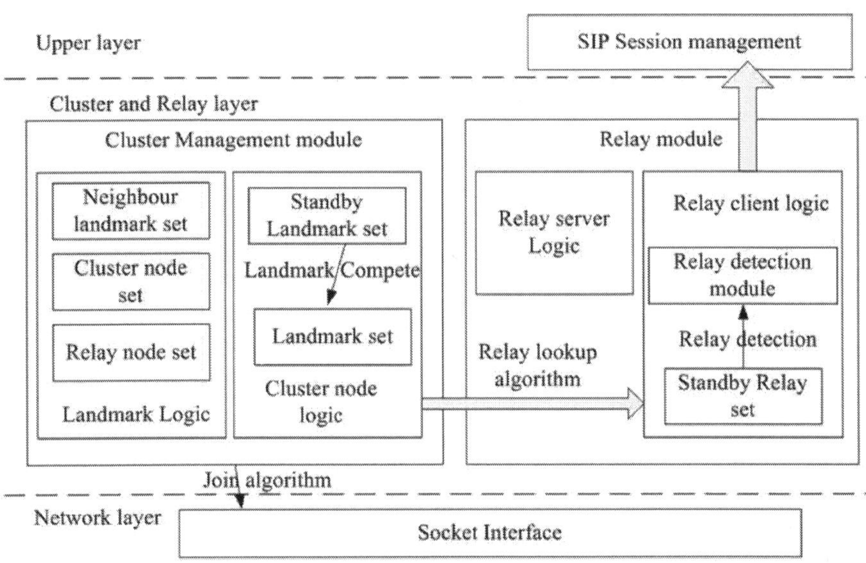


**Figure 3**. Module design.

The maintenance of topology: After a node joins a cluster, it will request near landmark list from its cluster landmark and detect them. If there is a landmark that is nearer than the current one, this node will join the new cluster by attaching to that landmark. This situation is common in mobile network because of node migration. In addition, each landmark stores a list of nearby ones. At the same time these landmarks will exchange their stored nearby landmark lists with each other so that the new joined landmark can be broadcasted in the network in time. Meanwhile, the landmark will also notify messages of new landmarks to its cluster nodes simultaneously. Then the nodes in the cluster will check their own condition and do some adjustments.

The maintenance of cluster node: the landmark maintains heartbeats with its cluster nodes and also with the relay nodes it stores.

Collecting information of relay: when a normal node joins the network, it will check out whether it could be a relay or not. If a relay node knows more than two landmarks whose path qualities (from itself to one landmark) are greater than a constant $Q_0$ value, it will notify the landmark. The landmark will also send back the list of previous landmarks addresses. The structure is called relay-cluster data structure in the system.

The initial communication: when two nodes communicate and need relay, they will first test the quality of the direct connection. And then, they will begin new inquiry for relays. Our SIP+P2P system supports multiple paths in one session, and uses optimizing routing [2,6] to improve QoS performance.

Inquiry for Relay: a normal node will ask landmark for candidate relay nodes by notifying the destination node's cluster, otherwise it will find candidate relay nodes locally. The candidate relay nodes are found in Relay cluster date structure. The landmark first searches the relay-cluster data structure for available relays of the corresponding destination cluster's relay nodes set and then sends the result to the source peer for detecting. If there are no proper relay nodes at last, the source node will ask for the near cluster set (which is got from the neighbor landmark set) and make an intersection with the destination node's near cluster set. Then it asks the cluster's landmark in the cluster set for relay nodes, makes them for the initial candidate relay nodes, and begins the detecting. If both of the two methods fail, the source node will randomly search nodes in its neighbors set for candidate relay nodes. Of course, this is the worst condition.

Building the communication: whether the source node gets the candidate relay list from local or from the landmark, these relays must be negotiated with the destination node, and then be detected. After the negotiation and detection, those suitable relay nodes will provide the application layer routing service.

## 3.4. Summary

Our system is a typical distributed system. As depicted above, the bootstrap server in our method is similar to a central server, and the landmark is similar to local server in CDN network. The cluster node should register and log off at the landmark and maintain a certain link with the landmark through heartbeat packages. All the landmarks make up a coverage network through the network topology, and maintain the cluster topology by heartbeat detecting. The landmark node and the cluster node compose a relation of tree. Landmarks exchange information with each other so that they could keep a close watch on the change of the network. And the landmark also maintains a list of relay of its autonomy cluster. The whole cluster overlay is similar to the Internet topology. The communication nodes request relays from its landmark and do the detection. If it fails, it will get the intersection of landmarks of the source node and the destination node. These landmarks and the relays in their clusters will be the candidate relays. This will reduce the load of the bootstrap server greatly and make the network very scalable. The node detection and relay detection are finished when building the coverage network topology, so the cost is shared with the process of building the coverage network topology. And this mechanism has a high hit ratio in the simulation system.

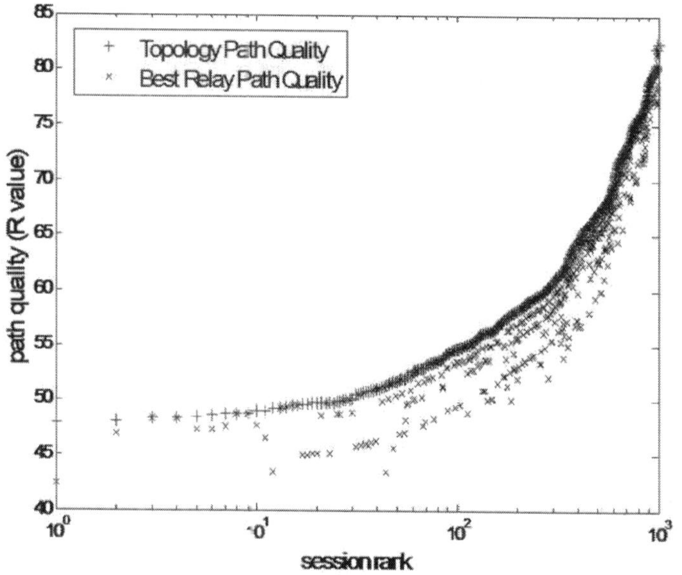


**Figure 4**. Topology path quality vs best relay path quality.

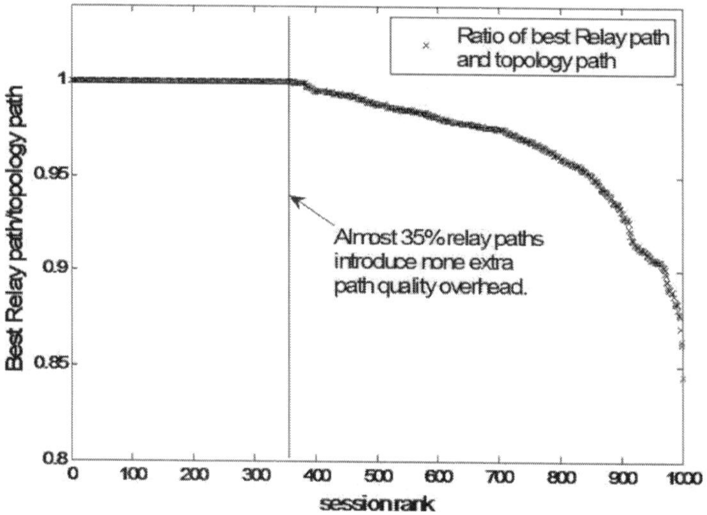


**Figure 5**. Distribution of relative quality overhead.

# 4. EXPERIMENTS AND ANALYSES

In this section, we will first introduce our experiment environments, including the simulation platform and the topology generator. Then we will introduce our evaluate methods. We mainly consider the performance of cluster overlay and the efficiency of relay lookup algorithm.

We evaluate our cluster overlay mainly from two aspects:

1) Firstly, we evaluate the similarity degree of topology between our cluster overlay and the Internet.

2) Then, we build cluster overlay in different network scales to evaluate its scalability.

The simulation results show that our cluster overlay topology is similar to Internet and has high scalability.

When evaluate the performance of relay lookup scheme, we mainly consider two evaluations below:

1) Hit ratio in different network scales.

2) Comparing path quality of our scheme with original SIP+P2P system and a simple random selection.

The simulation results show that our scheme has higher hit ratio and performs better than other schemes.

## 4.1. Experiment Environments

We simulate our design in the NS2 simulation platform. There are three main steps in our experiment:

1) Firstly, we use a topology generator tool to generate network topology which is as same as Internet.

2) Secondly, we build our cluster overlay based on this network topology.

3) Finally, we run relay lookup algorithms on our cluster overlay, and evaluate its performance.

In the NS2 environment, there is a tool to build network topology called GT-ITM [9]. We adopt the network model of random connecting of Transit AS with Stub AS (Transit-Stub model), which is most similar to the real network. With different sets of delay and packet loss between nodes (read from a static configure file), we guarantee that the topology is the same in different experiments for a given network scale. We build cluster overlay networks in different network scales to evaluate the scalability, and compare node pair's quality in cluster overlay with quality in original topology. After building cluster overlay, we run different relay lookup scheme to evaluate the performance of our topology aware algorithm.

## 4.2. Similarity Degree of Topology

We define relative quality ratio as the quality ratio of best relay path and topology path. When the relative quality ratio is higher, our cluster overlay topology is similar to Internet. We select 1000 node pairs and compute there topology path qualities and the best relay path qualities, which are plot in **Figure 4**.

Then, we compute the relative quality ratio, and plot its distribution below. As we can see in **Figure 5**, almost 35% relay paths introduce none extra quality overhead and 90% of relative quality ratio are above 0.9. These results prove our cluster overlay is very similar with real Internet topology.

## 4.3. Scalability of Cluster Topology

The scalability of this system requires that when the network scale increases greatly, the cluster scale does not increase greatly. The scalability is very important to our cluster overlay, because when the cluster overlay has high scalability, the maintenance cost of cluster topology does not increase greatly according to the network scale. As show in **Figure 6**, if the network scale increases, the number of landmarks and cluster nodes is increasing as well. But, the number of cluster nodes is increasing greatly and the number of landmark has no clear changes. This result proves that our scheme has good scalability.

## 4.4. Hit Ratio of the Best Relay Nodes

The object of this scheme is to find the best relays in P2P relay network which are near to both the source node and the destination node. This proportion of best relay found in this scheme is called hit ratio. In this experiment, we compare the relay nodes resulted from simulation with those resulted from calculation according to topology graph, and get a hit ratio. As show in **Figure 7**, we compare the hit ratio in different network scales for SIP+P2P and our new scheme. The results show that our scheme has better hit ratio than original SIP+P2P system. The reason is that our new scheme uses topology-aware to help look up the best relay.

## 4.5. Path Quality of Peer Relay Routing

In this section, we evaluate the QoS performance of our new scheme. We choose a network with 3000 AS nodes for simulation, and randomly select $10^4$ pairs of nodes sessions need relay. We only select those node pairs which are in different ASs, and whose path qualities are bad. Then we use different scheme to select relays and compare their path quality (Random selection, One-side optimal selection, Topology Aware, and Optimal search calculated according to topology graph):

As we can see in **Figure 8**, our new scheme performs better than One-side optimal and random selection, and is very close to optimal search. However, when using topology-aware lookup algorithm, there are 8% sessions can-not find relay. As mentioned above, when a client lookup relay failed, it will use a relay in local cache.

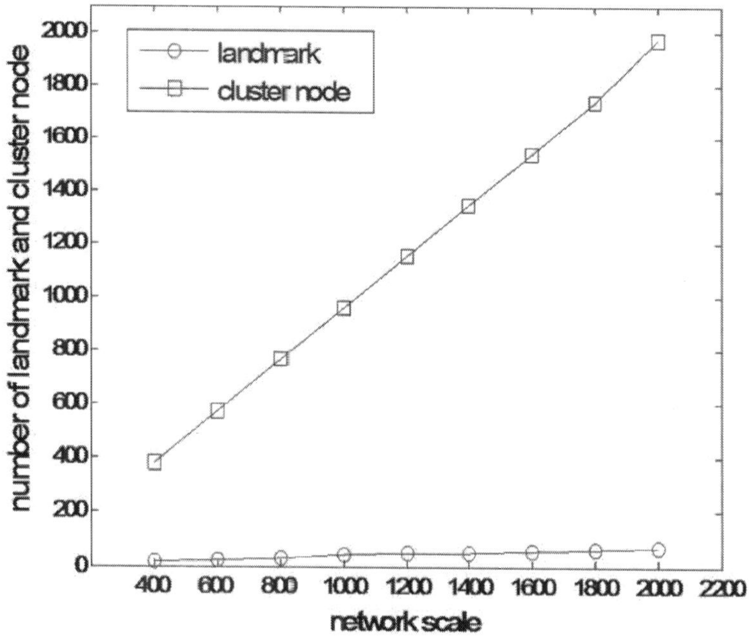


**Figure 6**. The number of cluster node and landmark.

## 5. CONCLUSIONS

In this paper, we propose a novel peer relay lookup scheme based on SIP+P2P system to improve its relay selection. This approach uses topology-aware technology to optimize the relay selection. First, we divide the whole P2P relay network into different clusters. When building the cluster topology, we use path quality between two nodes as the measurement metric. Then, when communicating node needs relay, it asks its landmark for proper relays and the cluster landmark will choose candidate relays according to the network topology. This new scheme can guarantee that the selected relay is near to both the source node and the destination node, which is better than the original SIP+P2P system.

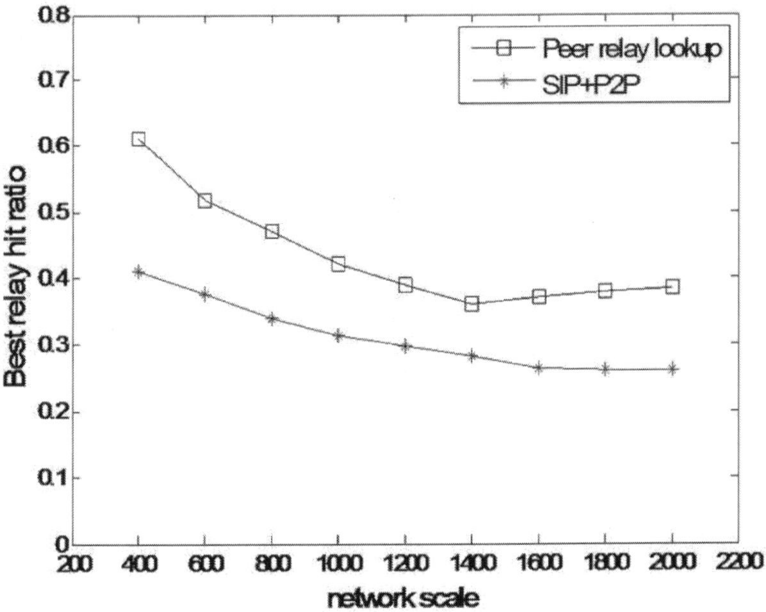


**Figure 7**. Hit ratio of the best relay.

We introduce the whole design in detail, including the system architecture, the peer design, and the node operation procedure. Then we evaluate our design in NS2 simulation, and the results show that our scheme is scalable. We also make comparison with traditional schemes and the results show that our scheme can get high best relay hit ratio and better path quality. Those experimental results confirm the feasibility of a real system.

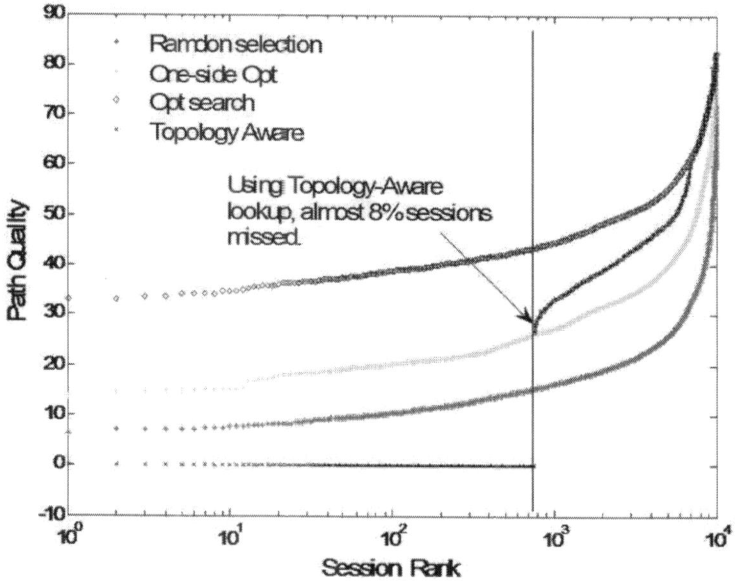


**Figure 8**. Path quality of peer relay routing

Our future work is to realize this scheme in our SIP+P2P UA [2], and to evaluate it in a real network environment. Moreover, to modify this scheme to make it suitable for mobile network is another future work.

## REFERENCES


1. S. Ren, L. Guo, and X. Zhang, "ASAP: An as-aware peer-relay protocol for high quality VoIP," Proceedings of 2006 IEEE ICDCS, pp. 70–79, 2006.

2. X. W. Zhang, W. M. Lei, and W. Zhang, "Using P2P network to transmit media stream in SIP-based system," Proceedings of ICYCS 2008, pp. 362–367, November 2008.

3. S. Guha and N. Daswani, "An experimental study of the skype peer-to-peer VoIP system," Proceedings of 2006 IPTPS, pp. 677–686, 2006.

4. D. G. Andersen, H. Balakrishnan, M. F. Kaashoek, and R. Morris, "Resilient overlay networks," ACM SIGCOMM Computer Communication Review, Vol. 32, No. 1, pp. 75–83, 2001.

5. D. G. Andersen, H. Balakrishnan, M. F. Kaashoek, and R. Morris, "The case for resilient overlay networks," Proceedings of 2001 Workshop on Hot Topics in Operating Systems, pp. 131–143, 2001.

6. S. Tao, K. Xu, A. Estepa, T. Fei, et al., "Improving VoIP quality through path switching," Proceedings of 2005 IEEE INFOCOM, pp. 515–523, 2005.

7. K. Suh, F. Daniel, K. Jim, and D. Towsley, "Characterizing and detecting skype-relayed traffic," Proceedings of 2006 IEEE INFOCOM, pp. 292–301, 2006.

8. H. Xie and Y. Yang, "A measurementbased study of the skype peer-to-peer VoIP performance," Proceedings of 2007 IPTPS, 2007.

9. K. Calvert and E. W. Zegura, "GT-ITM: Georgia tech internetwork topology models," In http://www.cc.gatech.edu/projects/gtitm/, 1997.

4

# AN EFFICIENT SCHEDULING SCHEME TO ENHANCE THE CAPACITY OF VOIP SERVICES IN EVOLVED UTRA UPLINK

**Yong-Seok Kim**

Telecommunication R&D Center, Telecommunication Network Business, Samsung Electronics, 416, Maetan-3dong, Yeongtong-gu, Suwon-si, Gyeonggi-do, 443-742, South Korea

## ABSTRACT

An efficient scheduling scheme is proposed to increase the available capacity of VoIP services over evolved UTRA uplink. On top of the advantages of persistent scheduling, the proposed scheme adaptively share the resources of two VoIP users to get early-termination gain of dynamic scheduler. Through system-level simulations, the performance of the proposed algorithm is evaluated in terms of the capacity enhancement of VoIP services. Comparisons with the original persistent scheduling and the HSUPA scheduler reveal that the proposed scheme increases the capacity of VoIP services up to 20%.

# 1. INTRODUCTION

Evolved universal terrestrial radio access (E-UTRA), which is known as long-term evolution (LTE) of third-generation cellular system, is being specified by the third generation partnership project (3GPP). In September 2006, the study item of the LTE has been completed and the corresponding work item was scheduled to be finalized within about one and a half years, that is the second half of 2007, so that the subsequent initial deployment can be possible in the year of 2009 or 2010. The E-UTRA is regarded as the preliminary version of next generation wireless communication system because of its capability to satisfy demand for higher user bit rates [1, 2]. In order to obtain such higher user bit rates, the E-UTRA is being designed by only packet-switched (PS) network without circuit mode, requiring that all the available LTE services should be implemented on top of internet protocol (IP). At this point, the transmission of real-time data such as voice traffic through PS IP network becomes arguably the hottest issue today because voice over IP (VoIP) has high visibility in consumer space.

Managing such real-time data transmission, scheduling algorithm at medium access control layer can be a core function because the algorithm directly controls the level of quality-of-service (QoS). As a basic scheduling scheme for packet-based services, proportional fairness (PF) scheduler was designed to support the high data rate of 3GPP2 wireless system in [3]. The PF scheduler provides effectiveness from the view point of throughput and fairness by judiciously selecting frames based on the average and the instantaneous data rate of each user. Because of its simple yet effective scheduling capability, the scheme is still being used in the LTE as dynamic scheduler. However, such dynamic scheduler could provide limited performance for some delay sensitive real-time services in that the scheduling algorithm passed over delay constraint in its frame prioritization.

To cope with such delay problem and to satisfy the specific QoS parameter of maximum allowable service delay, the author proposed a frame bundling scheme in [4]. The scheme modified PF scheduler such that the estimated delay of each user controls the priority of frame schedule with frame bundling according to the user's channel condition, resulting in the significant enhancement of the capacity of VoIP over high-speed downlink packet access network. Nevertheless, this work did not reflect on the overhead of control signaling, although it grows fast as the number of VoIP users increases. Note that the control signaling is required to assign and distribute resources for each user at every transmission time interval (TTI), and the TTI is usually very short, that is, 1 millisecond in most of the emerging systems. This overhead may degrade the spectral efficiency of radio systems seriously, making its minimization essential to the enhancement of systems performance.

As a method to reduce such control signaling overhead, persistent scheduling scheme has been investigated in [5, 6]. By use of the inherent characteristics of voice traffic such as frame size and period, the scheme efficiently reduced control channel signaling overhead. With the aid of such overhead reduction, the persistent scheduling scheme has been discussed as an option for VoIP services in E-UTRA uplink. However, such persistent resource allocation makes lack of early-termination gain, bringing about the waste of frequency resources with the reduced fairness of users. This is because the allocated resources through this persistent scheduling scheme (i.e., the TTI and frequency resource block (RB)-index) are assigned to each VoIP user for a relatively long period of time without changes. Accordingly, the persistent scheduling scheme could limit the capacity of VoIP service using LTE system, and thus more efficient scheduling should be required.

In this paper, an efficient LTE scheduler is proposed to increase the capacity of VoIP in E-UTRA uplink. The proposed scheme modified the persistent scheduling algorithm such that the resources of two VoIP users can be coupled. By letting these coupled resources adaptively shared by the two VoIP users, the proposed scheme achieves a significant amount of early-termination gain without the need of additional control signals.

The remainder of this paper is organized as follows. In the next section, the conventional and the proposed scheduling schemes of E-UTRA uplink system will be described. Then, Section 3 presents the details of simulation configurations and the criteria to measure the capacity of VoIP services. Section 4 follows to discuss the performance of the proposed scheme by system-level simulations. In this section, we also present the comparison results of the proposed algorithm with the original persistent scheduler of LTE as well as that of high-speed uplink packet access (HSUPA) in 3GPP Release'6 [7], which is the latest version of already deployed wireless network, for more thorough comparison. Finally, the conclusion of this paper is drawn in Section 5.

# 2. SCHEDULING FOR VOIP SERVICES IN E-UTRA UPLINK

This section deals with the question of *what kind of resource allocation scheme is suitable* for satisfying the required QoS of various services having different transmission characteristics. Those services can be voice, streaming, web-browsing, file transfer, and so on.

## 2.1. Conventional Dynamic Scheduling

Basically, dynamic resource allocation has been chosen for integrated scheduler of E-UTRA uplink transmission scheme. The integrated scheduler includes packet scheduler, adaptive modulation and coding (AMC) unit, hybrid automatic repeat request (HARQ) manager, power management unit, and buffers [8]. They are all located at eNB to support fast channel-dependent scheduling.

The packet dynamic scheduler is the main and the representative part of such integrated scheduler to select users for the assignment of time/frequency resources at every TTI. Resource block (RB), which consists of 12 subcarriers, is the minimum scheduling granule of such packet scheduler in frequency-domain. With the unit of RB, in E-UTRA uplink, control parameters such as payload sizes and modulation coding sets (MCS) are determined by the eNB's packet dynamic scheduler, depending on queue states in the user's data buffer. Therefore, the reporting of the buffer status to eNB is essential. This allows for all the tight QoS control taken by the eNB and no QoS handling done at user-site.

The algorithm of packet scheduling can be divided into two parts, saying the selection of users and the assignment of RBs. In the users selection procedure, dynamic scheduler chooses a particular user according to his/her priority that is calculated based on the PF algorithm. As for the RBs assignment, the scheduler exploits the benefit of multiuser frequency diversity. To do so, the RBs assignment for selected retransmission is done by picking up relatively good quality of channels for the previous TTI. However, once RBs were allocated to a user, the retransmission packets of HARQ uplink should be maintained until the packets are correctly received or the maximum allowable time interval. This is because the E-UTRA system employed synchronous HARQ and nonadaptive AMC scheme for uplink to reduce the amount of control signals. Due to this limitation, it becomes less flexible to assign good RBs for the retransmission of uplink data, but HARQ combining gain still can benefit the retransmission quality.

However, downlink control channel signaling is classified into two groups to support uplink services. The first group is used to indicate the assigned RB's information. It contains some user specific signals such as user identification (UEID), cyclic redundancy check (CRC), RB assigned frequency-domain location, assignment time-domain duration, modulation, coding, and payload size [2]. Also, this control channel signaling is required at the beginning of each first transmission TTI because of the synchronous HARQ in E-UTRA uplink. The second is used to assign RBs in order to apply ACK/NACK information for uplink data.

## 2.2. Original Persistent Scheduling

Since the eNB handles a large portion of VoIP users, the amount of the required control channel signals for voice services increases significantly. It is critical that the control channel signaling overhead is minimized as much as possible. Therefore, for voice services, this paper focuses on the persistent resource assignment with the static control signaling. The persistent scheduling is mainly aimed for the voice services. It has been discussed for E-UTRA as a solution to overcome the limitation of layer1/layer2 (L1/L2) control channels.

By definition, the persistent assignment means that the resources are assigned by the eNB's persistent grant for a relatively long period of time (i.e., talk-spurt period). Once an allocation it is not changed regardless of channel quality and queue status except when entering silent period. It is not required to inform the scheduling L1/L2 control channel signals except for the first establishing time [5, 6]. So it is very efficient for the control signaling overhead reduction. The eNB's scheduler can simply use the predefined persistent RBs in the allocated every TTI. The assigned resources are released only when the VoIP user at the talk-spurt state enters the silent state. During the silent period, the silent payloads are transmitted by using the conventional dynamic scheduling method. This is because the L1/L2 control channel signaling overheads are not burden since the period of the silent frame is more than talk-spurt frame (i.e., 160 milliseconds).

An example for the resource assignment scenario according to the original persistent scheduling is illustrated in Figure 1. As you can see the figure, the frequency-domain position and the transmission TTIs of the assigned resources are statically configured. Also, these configurations are repeated in every repeated pattern period. However, these persistent assignments are once coming on when a voice radio bearer is established, and then the configurations are not changed until the state transition. Therefore, the L1/L2 control channel signals are almost not needed in the persistent scheduling scheme.

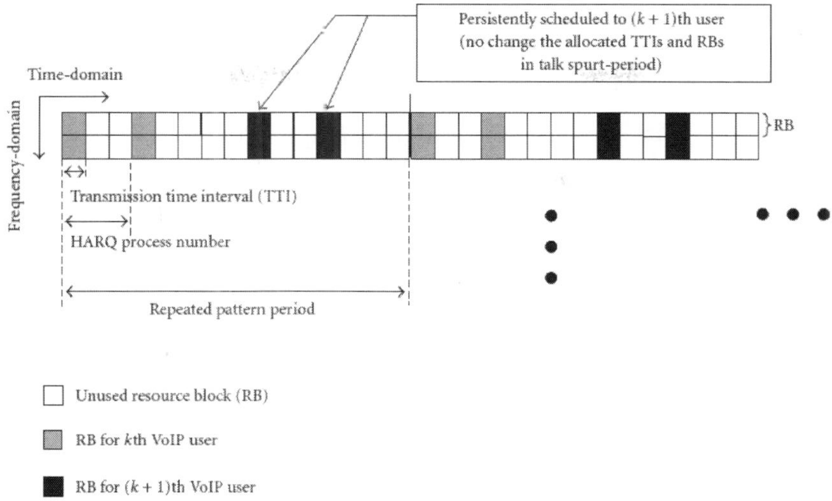


**Figure 1.** Resource assignment scenario according to the original persistent scheduling.

Although the persistent scheduling is very efficient to reduce the overheads of L1/L2 control channel signals, it is very inefficient to achieve some dynamic scheduling benefits such as early-termination gain and channel-dependent scheduling. The early-termination gain means that the allocated resources, which are assigned to a scheduled user, can be potentially available to other user when they are not used by the allocated user. It may be efficient in resource utilization. But, in the original persistent scheduling scheme, the allocated resources are not changed during the talk-spurt period. In other words, since the original persistent scheduling does not provide the early-termination gain, the available capacity of VoIP is limited. Therefore, there is a need for a solution to improve the available VoIP capacity.

## 2.3. A Proposed Scheduling Scheme

### 2.3.1. Concept of the Proposed Scheduling

In this section, an adaptive resource sharing scheme using user pairing, which can maintain the properties of the original persistent scheduling method, is proposed. Design objective of the proposed scheme is to achieve the early-termination gain by using user pairing of two VoIP users.

An example of the resource assignment scenario with the resource sharing through user pairing is presented in Figure 2. We note from the figure that the RBs are allocated persistently by a unit of the paired user group. This way is different from the original persistent scheduling. Moreover, in each pairing group the authority of the transmitting voice packet through the shared RBs is adjusted adaptively. It is applied by fitting into the event status, which is illustrated in Table 1. Unlike original persistent resource assignment scheme, the authority of the shared RBs can be changed adaptively in each predefined period (i.e., repeated pattern period). In other words, the number of the allocated transmission TTIs can be changed adaptively in accordance with each paired user's packet error at the previous period. Therefore, we can call that the proposed scheduling scheme is a channel-adaptive scheduling method. The repeated pattern period can be set depending on the arrival and discard time of voice packet. Each newly arrived voice packet in the user's data buffer is associated with the packet discard time. The scheduler specifies the maximum time in which an arrived packet is allowed to be buffered in the user data buffer before it should be dropped. In this paper, the repeated pattern period is set by 20 milliseconds, which is the same as the packet arrival time interval. More precisely, the data transmission opportunity for the 1st pairing user in $k$ th pairing group increases when the event 3 status occurs.

**Table 1.** Event status for indicating the authority of a shared RB.

| Number | 1st pairing user's ACK channel | 2nd pairing user's ACK channel | Status |
|---|---|---|---|
| 1 | ACK | ACK | Maintaining |
| 2 | ACK | NACK | Changing |
| 3 | NACK | ACK | Changing |
| 4 | NACK | NACK | Maintaining |

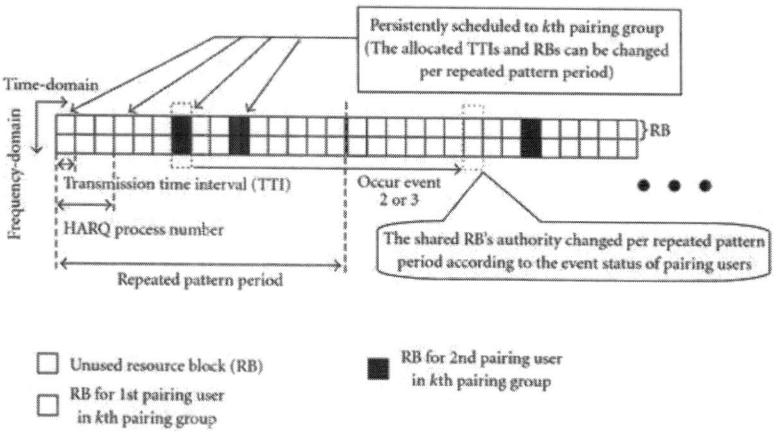


**Figure 2.** Resource assignment scenario with a resource sharing through user pairing.

In order to perform the above process, one paired VoIP user in each pairing group only monitors the other paired user's ACK/NACK channel information. So, there is no burden for control channel signaling to change the transmitting authority adaptively. However, the rest resources, except for the shared resources for pairing groups, are used according to the original persistent scheduling rule. Moreover, if the only one user in a pairing group is staying alone because the other paired user's assigned resources are released, the remaining user can be operated according to the original persistent configurations.

### 2.3.2. User Pairing Method

The resource sharing algorithm that we propose for VoIP services in E-UTRA uplink is designed to obtain the early-termination gain. It can be achieved by employing the user pairing. The user pairing can be classified as two methods basically. One is a random pairing that two supported VoIP users can be paired as a pairing group without any consideration. It does not take into account the individual user radio channel condition, the data buffer status, and the packet transmission delay in making a paired group. The other is a best pairing scheme which pairs two VoIP users to a paired group considering the individual status of user conditions. In this scheme, the best pairing chooses one user under good conditions and the rest under bad status of conditions in making a paired group. In other words, the delay sensitive user, who requires more retransmission opportunities due to a bad channel conditions, is favored by pairing a group with the delay nonsensitive users under a good channel condition. This pairing configuration can increase the number of event 2 or 3 occurrence. So, the discarded packets, which occur because of a timeout of the delay sensitive user, can be reduced by using the shared resources of the delay nonsensitive user. Therefore, best pairing can give a large efficiency to the proposed resource sharing scheduling scheme. However, in this paper the above two pairing methods are compared to each other.

### 2.3.3. Description of Overall Operation for a Proposed Scheduling Scheme

In this section, we describe the overall procedure of the proposed resource sharing scheduling scheme. The authority adaptation is executed every repeated pattern period. Also, the user pairing can be also adjusted several times of repeated pattern periods. The overall procedure can be described on the eNB-site and user-site, which is summarized as follow.

The Flow on the eNB-Site

a. pair two users according to the user pairing method;

b. assign RBs to each pairing group persistently using the measured priority that can be calculated by the modified PF in [4];

c. scheduler monitors the event status of the paired users in each pairing group;

d. change the transmission authority of shared RBs when they meet the boundary of repeated pattern period and occur event 2 or 3.

The Flow on the User-Site

a. monitor his own ACK/NACK channel as well as the ACK/NACK channel allocated to the rest user in a pairing group;

b. change the available transmission RBs when they meet the boundary of repeated pattern period and occur event 2 or 3;

c. transmit VoIP packet using the allocated RBs at his assigned time.

### 2.3.4. Discussion on the Control Channel Signaling Overhead

In this section, we investigate a quantitative comparison of two VoIP scheduling schemes from the control channel signaling overhead point of view. However, in the proposed resource sharing scheme each paired user must know the authority of the shared resources to perform according to the proposed scenario. Therefore, one concern for the proposed scheme is the additional control channel signals.

The paired user's control channel information for ACK/NACK channel needs to be once transmitted in the first time when the voice radio bearer is established. Therefore, the achievable capacity gains using the proposed scheduling are enough even if the additional control channel signaling overhead is taken into account, since the additional control channel signals are a very small quantity.

### 2.3.5. Power Control and Interference over Thermal in Uplink

The focus of this section is on slow power control and interference over thermal (IoT). The IoT indicates the ratio between the total received interference power and the thermal noise. It is similar to the rise-over-thermal (RoT) in HSUPA. The IoT is computed per sector per subframe and is defined as [9]

$$\mathrm{IoT} = \sum_{i;\mathrm{eNB}(i)\neq s} \frac{E_{i,s} + N_0}{N_0}, \quad (1)$$

where $\mathrm{eNB}(i)$ represents the serving sector of user $i$ , $E_{i,s}$ is the total received power from user $i$ at sector $s$ , and $N_0$is the thermal noise variance. The IoT plays an important role in determining the system throughput since the transmission rate depends on the signal-to-interference plus noise ratio (SINR) and not just on the signal-to-noise ratio (SNR). The IoT bounds the additional interference to the cell, and thereby limits the power required for new users to access the network. Typical IoT values in commercial networks range from 3 to 10 dB and we will employ the worst constraints the same as or near the average 3 dB in this paper [9].

However, similar to HSUPA, the average IoT level received at the eNB is effectively reduced by using uplink power control approach. To achieve good tradeoff between the cell-edge performance and the overall spectral efficiency, power control scheme should be considered carefully. The only inter-cell interference exists in the uplink for E-UTRA due to an orthogonal access. Therefore, the slow power control is sufficient to compensate the influence of IoT. In this paper, the slow power control may be implemented in each eNB by sending power control command slowly. It should be noted that the other approach is possible. For example, each user can derive its own transmit power based on the path loss measurement from downlink pilot.

# 3. VOIP SERVICES OVER E-UTRA UPLINK

## 3.1. The Properties of VoIP Services

### 3.1.1. Traffic Model and Protocol

In this paper, the conversation traffic can be approximated to the two state Markov model with a suitable voice activity factor (VAF) [10]. The adaptive multirate (AMR) voice codec is mandatory for voice services in E-UTRA systems. The AMR codec generates a 32-bytes voice payload every 20 milliseconds during talk-spurt period. During silent period, a 7-bytes payload carries a silence descriptor (SID) frame every 160 milliseconds [11]. Also, we can set by 50% VAF.

However, a typical VoIP protocol stack, which employs the real-time transport protocol (RTP), is encapsulated to the user datagram protocol (UDP). This, in turn, is carried by IP. These combined protocols demand a 40-bytes IPv4 header or a 60-bytes IPv6 header. Therefore, the overhead, caused from these headers, seriously degrades the spectral efficiency in supporting VoIP service. Therefore, efficient and robust header compression (ROHC) technique must be used to reduce the amount of the large headers in the IP/UDP/RTP layers. This technique can reduce the size of the IP/UDP/RTP headers as little as 2 or 4 bytes using IETF RFC 3059 [12, 13]. Therefore, in this paper we assume that the IP/UDP/RTP headers are reduced as 4 bytes using ROHC.

### 3.1.2. Definition of VoIP Capacity

In PS network, packets will be dropped due to packet error and packet delay exceeding the target latency. Although some packet loss occurs, the voice quality is not affected if the amount of packet loss is less than outage threshold. At this point, the VoIP capacity is defined by the maximum number of VoIP users that can be supported without exceeding a given outage threshold. The outage criterion means that packet error rate (PER) of VoIP user is kept within 2%. Moreover, at least 95% of total VoIP users should meet the above outage criteria [12].

### 3.1.3. End-to-End Delay Latency for QoS Support

To ensure end-to-end QoS, the low delay is one of the most important factor for maintaining high-quality VoIP services. But, to achieve high VoIP capacity, the scheduler must have sufficient time to manage voice packets. Of the assumed 200 milliseconds mouth-to-ear delay latency for qualified voice services, about 50 milliseconds are available for scheduling in the uplink [12]. The delay in others such as processing time for downlink/uplink, scheduling time for downlink, IP, and backhaul network delay is bounded to 150 milliseconds. It is the fixed value that allows us to focus on the delay within radio access network as shown in Figure 3. Although VoIP performance depends on both downlink and uplink performances, we would like to set aside the consideration of both directions as a comprehensive study for future research.

## 3.2. System-Level Simulation Setup

### 3.2.1. Physical Layer in E-UTRA Uplink

For uplink transmission, the important property is to allow user equipment (UE) for power efficient transmission to maximize coverage. The choice of single-carrier frequency-domain multiple access (SC-FDMA) is therefore preferable in E-UTRA uplink. This is because the resulting peak-to-average power ratio (PAPR) is lower than orthogonal frequency division multiple access (OFDMA) in downlink. Also, the fast fourier transform (FFT) and inverse fast fourier transform (IFFT) are used in transmitter to produce the FDMA signal. So, multipath propagations are handled by frequency domain equalization at the eNB, aided by the insertion of a cyclic prefix (CP) in the transmitted signal. A subcarrier spacing of 15 kHz is adopted, which allows for simple implementation of dual mode between wideband code division multiple access (WCDMA) and E-UTRA terminal. Also, the number of individual subcarrier as 12 consists of 1 RB. The RB is the basic time-frequency transmission resource unit in E-UTRA system. To minimize delays, the subframe duration, that is TTI, is selected as short as 1 millisecond, corresponding to 14 OFDM symbols [2].

However, in case of scheduled access the eNB-driven scheduling scheme has been chosen in uplink. It means that unlike HSUPA, the uplink data transmission format of a UE such as payload size and MCS level is controlled by eNB.

### 3.2.2. Simulation Environments

To investigate the capacity performance of VoIP traffic, the system-level Monte-Carlo computer simulation is accomplished in this paper. In all simulations, 5 MHz system bandwidth is considered among the flexible bandwidths. The simulations are carried out with a regular hexagonal 19 cellular model, where the intersite distance (ISD) between eNB is 500 m. Mobile terminals should be uniformly distributed on the 19-cell layout for each simulation run and assigned by typical urban (TU) channel model according to channel assignment probability specified in [2]. Note that a realistic model of the wave propagation plays an important role for the significance of the simulation results. Mobile speed is 3 km/h since 3GPP E-UTRA system should be optimized for low mobile speed. Shadowing is modeled by log-normal fading of the total received power. An attenuation is determined by Hata model.

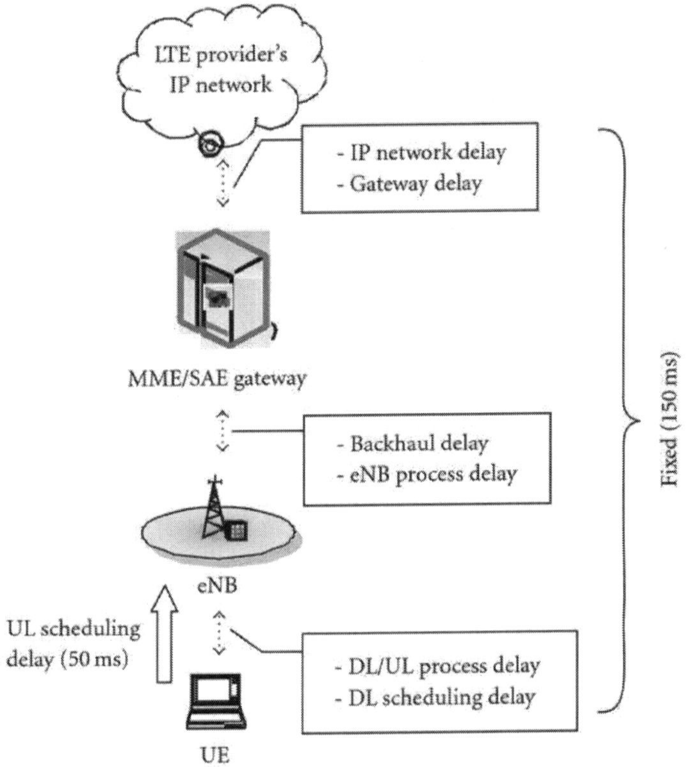


**Figure 3.** End-to-end delay components in E-UTRA.

Moreover, we consider resources for control and common channel overheads such as demodulation (DM) reference signal (RS) and uplink ACK/NACK chancel for downlink transmitted data [14]. The synchronous HARQ module with 8 process numbers is employed in this paper. Also, slow power control for uplink is employed as 50 Hz. The applied MCS level is fixed as QPSK and 0.55 coding rate for VoIP services. Finally, we simulate 40 000 TTI snapshots (40 seconds) in average for investigating the performance of system. The main simulation parameters are summarized in Table 2.

**Table 2.** System-level simulation parameters.

| Parameter | Assumption |
|---|---|
| Source traffic packet overhead | AMR 12.2 kbps,<br>VAF = 0.5, 2-state Markov,<br>ROHC 4 bytes [IETF RFC 3059] |
| Cellular layout | Hexagonal grid, 19 sites, 3 sectors<br>(eNB-to-eNB 0.5 km)<br>Carrier frequency 2 GHz |
| System bandwidth | 5 MHz |
| Propagation loss | Path loss = −128.1–37.6*log(R) |
| Shadowing model | Log normal Std. dev. 8 dB,<br>[Hata model] |
| User speed | 3 km/h |
| Antenna gain | Node B 14 dB/UE 0 dB |
| Penetration loss | −20 dB |
| User Tx. Max. power | 24 dBm |
| Fading model | Typical urban (TU) |
| Thermal noise density | −174 dBm/Hz |
| Number Rx ant. | 2 |

| | |
|---|---|
| Retransmission | No RLC retransmission (No ARQ),<br>sync. HARQ (max. retrial = 4),<br>HARQ process number = 8,<br>chase-combining |
| Power control | 50 Hz (sounding RS-based closed-loop method) |
| Scheduling | Persistent scheduling with a resource sharing using user pairing,<br>(repeated pattern period = 20 milliseconds) |
| MCS level | QPSK, 0.55 coding rate,<br>talk-spurt period (2 RBs assign);<br>silent period (1 RB assign) |
| Link curve mapping | Effective SIR method (ESM) |
| Link curve TTI | 1 millisecond |
| Considered overheads | 29%(pilot and control overheads)<br>2 long blocks for DM RS<br>4 RBs for control signals |

# 4. VOIP CAPACITY EVALUATION

In this section, we evaluate the available capacity of VoIP traffic with the proposed scheduling scheme in the typical urban fading channel environments. The proposed scheme is also compared with the original persistent method.

The percentage of VoIP users satisfying outage limitation (e.g., 2% PER) is presented in Figure 4. It is shown as a function of the available delay latency in scheduler. From the figure, we observe that the percentage of users satisfying outage criterion increases according to the increase of the available delay latency in scheduler. This is because the probability of packet loss because of a timeout of target delay latency becomes more decreased with higher delay latency in scheduler. In addition, a proposed scheme results in the smaller required delay latency in scheduler to achieve the same outage performance than the corresponding original persistent scheduling scheme. For example, If we aim for an identical percentage value of 0.8 and the same number of users as 190, the required delay latency using a proposed scheme is 43 milliseconds when employing a random user pairing and 39 milliseconds when employing a best user pairing, respectively. On the other hand, when using an original persistent scheme the required latency may be 52 milliseconds. This implies that the VoIP capacity can be increased if the proposed scheduling scheme is employed. Specifically, more capacity benefits can be obtained by employing a best user pairing approach.

Figure 5 characterizes the achievable VoIP capacity, when the delay latency in scheduler is 50 milliseconds statically. According to the figure, we note that the available VoIP capacity with a proposed scheme employing a best user pairing is 200 against that the 168 VoIP users can be serviced by using original persistent scheduling.

However, Figure 6 shows the IoT distributions for the different number of VoIP users. The IoT is computed per sector per subframe. Unlike downlink environments, in uplink case IoT level will be an important factor determining the VoIP capacity. We will employ the worst constraints the same as or below the average 3 dB in the paper.

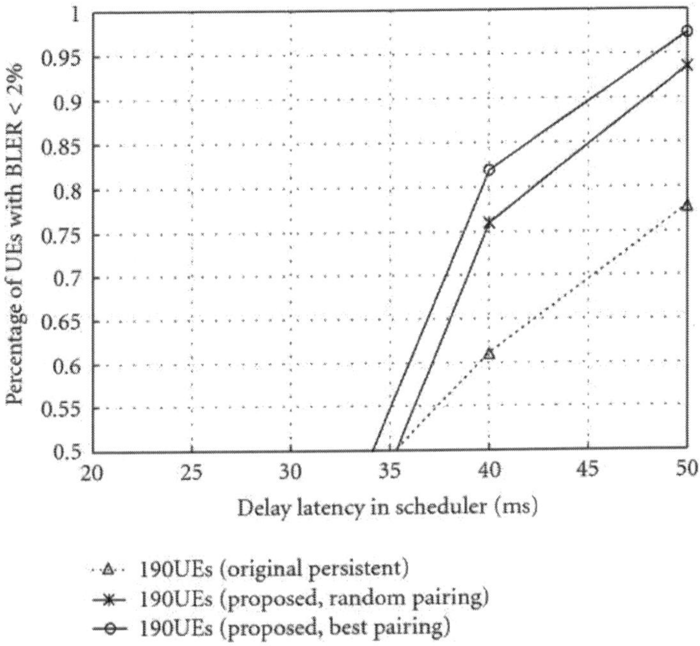


**Figure 4.** Outage probability versus delay latency in scheduler.

Finally, Table 3 summarizes the capacity of VoIP using proposed scheduling scheme in E-UTRA uplink. The results confirm that the VoIP capacity with a proposed scheme can be improved of 10% when using the random user pairing and 20% when using the best user pairing approach against the original persistent scheduling. Moreover, the available VoIP capacity employing a best user pairing scheme over E-UTRA uplink provides significantly highly capacity (e.g., 160%) if compared to the VoIP capacity over HSUPA (Release'6).

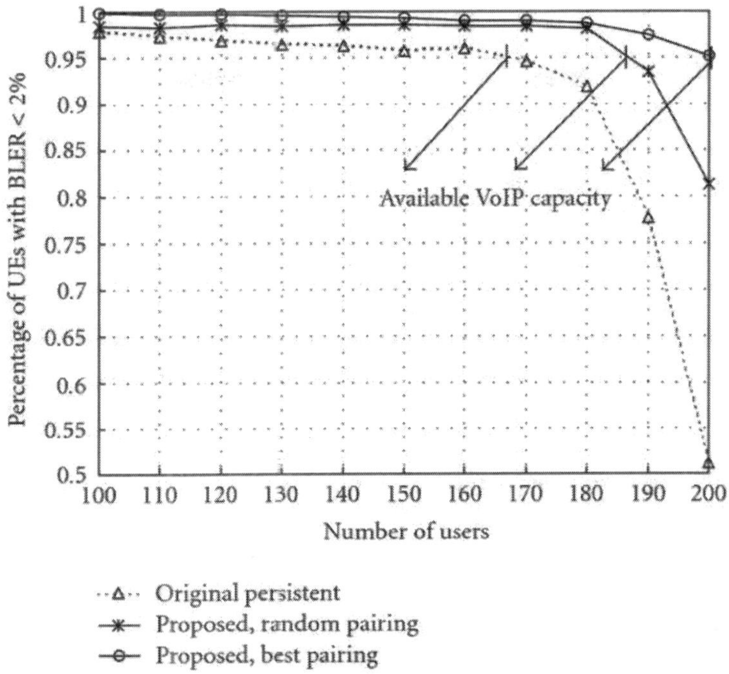


**Figure 5.** Outage probability versus number of VoIP users.

## 5. CONCLUSION

In this paper, we propose the efficient scheduling method employing a resource sharing approach. This proposed scheme employs the random user pairing and the best user pairing method to improve the capacity of VoIP services over E-UTRA uplink. Results are investigated by the system-level simulation. Our simulation results show that the employment of proposed scheduling scheme makes a larger available capacity than that resulted by the original persistent scheduling. In addition, we also conclude that E-UTRA is attractive for supporting of VoIP services if compared to HSUPA (Release'6).

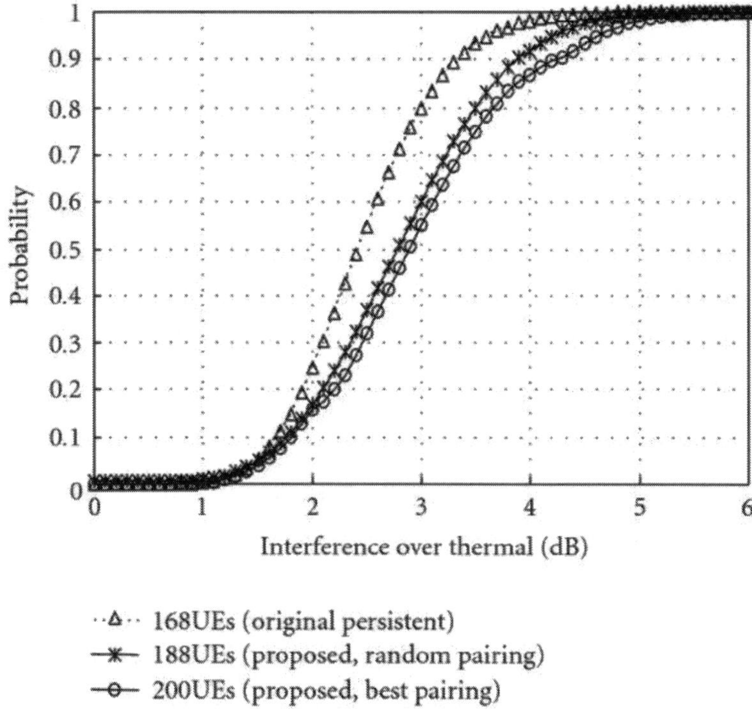


**Figure 6.** VoIP user IoT cumulative distributions.

**Table 3.** Summary of VoIP capacity using proposed scheduling scheme.

| | Capacity for VoIP services | IoT [dB] |
|---|---|---|
| Original persistent scheme | 168 | 3 |
| Proposed (random pairing) | 188 | 3.2 |
| Proposed (best pairing) | 200 | 3.3 |
| HSUPA (Release'6) | 70 | 3 |

The consideration of the combination of other traffic types such as best-effort, web, and streaming may be an interesting issue for future study. Moreover, the influence of ACK/NACK decoding errors may be considered.

## REFERENCES


1. 3GPP TR 25.913: Requirements for Evolved UTRA (E-UTRA) and Evolved UTRAN (E-UTRAN).

2. 3GPP TR 25.814: Physical layer aspect for evolved Universal Terrestrial Radio Access (UTRA).

3. Jalali A, Padovani R, Pankaj R: Data throughput of CDMA-HDR a high efficiency-high data rate personal communication wireless system. *Proceedings of the 51st IEEE Vehicular Technology Conference (VTC '00), May 2000, Tokyo, Japan* 3: 1854-1858.

4. Kim Y-S: Capacity of VoIP over HSDPA with frame bundling. *IEICE Transactions on Communications* 2006, E89-B(12):3450-3453.

5. 3GPP R2-061920 Persistent Scheduling

6. 3GPP R2-062788: Persistent scheduling and dynamic allocation.

7. Chen T, Kuusela M, Malkamaki E: Uplink capacity of VoIP on HSUPA. *Proceedings of the 63rd IEEE Vehicular Technology Conference (VTC '06), May 2006, Melbourne, Australia* 451-455.

8. Kim YH, Kim Y-S: A scheduler for multi-traffic services in WCDMA networks. *Proceedings of the International Symposium on Communications and Information Technologies (ISCIT '06), October-September 2006, Bangkok, Thailand* 1004-1007.

9. Hande P, Rangan S, Chiang M: Distributed uplink power control for optimal SIR assignment in cellular data networks.*Proceedings of the 25th IEEE International Conference on Computer Communications (INFOCOM '06), April 2006, Barcelona, Spain* 1-13.

10. 3GPP TR25.896: Feasibility study for enhanced uplink for UTRA FDD.

11. 3GPP TS 26.236: Packet switched conversational multimedia application—transport protocols.

12. 3GPP R1-070674: LTE physical layer frame work for performance verification.

13. IETF RFC 3059: Attribute list extension for the service location protocol. February 2001

14. The 3GPP TS 36.211: Physical channels and modulation. (Release 8), v100

5

# RE-ENGINEERING CAMPUS-WIDE INTERNET TELEPHONY USING VOICE OVER INTERNET PROTOCOL

**Francisca O. Oladipo, Christogonus C. Madu, Christopher C. Okoro**

Computer Science Department, Nnamdi Azikiwe University, Awka

## ABSTRACT

This work presented the re-engineering of a university's telephone system through the design of alternative implementation and specification scenarios for a campus-wide telephony system based on Voice over Internet Protocols (VoIP) technology; and demonstrated that the technology can be adapted for use in our university community to make cheaper calls using the desk Internet Protocol (IP) phone and data services. This research drew inspirations from similar efforts by some institutions to replace their old PSTN telephony system with the VoIP-based architecture. The Analysis phases of the Structured Systems Analysis and Design Methodology (SSADM) were adopted in carrying out a detailed study of the underlining technologies of VoIP and the possibilities of implementing a campus-wide telephony system using the technology; prototyping was deployed to build a prototype VoIP-based telephony system and Cisco Packet tracer was configured to run several simulation sessions of the developed specifications. The simulation results showed that VoIP can be successfully deployed to provide in a flexible manner, additional data-driven services in campus-wide telephony through a merger of telephone and information technology facilities.

## KEYWORDS

VOIP, Telephony, Prototyping, Cisco packet tracer, Design specification

## 1. INTRODUCTION AND BACKGROUND

The technology of Voice over Internet Protocol (VoIP) involves using the technology of the Internet Network to deliver voice communications and multimedia session as packets over the network. Because VoIP is a telephony service, it is also referred to as broadband phone service, internet telephony, IP telephony, and broadband telephony. Due to the affordances of Internet-ready phones, and the fact that the IP is the communication protocol of most devices, VoIP is best positioned to be the service platform for next-generation application [1]. The Public Switched Transfer Network (PSTN) on the other hand is a connection-oriented, circuit-switched network that uses dedicated channels for transmission. The PSTN had switched to transmitting digital signals to solve the problems associated with its original analog transmission using Pulse Code Modulation (PCM) to convert all analog signals into digital transmissions at the calling and receiving ends. However, the PSTN suffers two significant disadvantages: high cost resulting from the expensive bandwidth and an inefficient use of networking channels.

VoIP unlike its predecessor, the PSTN which is currently built on a closed infrastructure; is built on an open infrastructure and several vendors can provide applications and access. While the PSTN technology involves vendors only building applications specific for their equipment and its current architecture has not made it possible for many vendors to write new applications for it; VoIP allows the development of more creative solutions and applications as well as the convergence of data, voice and video in one channel.

This research is aimed at simulating a VoIP-based telephony service using a university communication system. We believe that if we can successfully implement the simulation, and given the advantages of VoIP over the traditional PSTN technology, a modern telephony system can be implemented for the campus. The researchers aimed at reducing cable clusters around offices, facilitating in-office communication and enabling conference calling at any point where the IP phone or VoIP phone is being installed. The project covered the interconnection of different offices and the researchers, with the help of simulation showed that it was possible to make cheaper calls using the desk IP phone and video streaming on the Soft Phones and interfaces installed on Personal Computers (PC).

## 2. RELATED RESEARCH

Reference [2] believed that the technology of VoIP has a great potential to lower the cost of telecommunication service and provide better Quality of service (QoS) due to its ability to merge both data and voice in a single channel. Their research conducted a performance analysis of the VoIP technology using evaluation criteria of QoS, Mean Opinion Score, packet loss ratio and jitter. This analysis was done over popular applications like Skype and msn. Five scenarios were covered: VoIP Call in LAN, Long Distance VoIP Calls under LAN, VoIP calls in LAN with ftp server, VoIP Calls in WLAN, and VoIP in WLAN with interference. Their result revealed that implementation over Ethernet exhibited more stability and less delay than wireless connection and that interference near wireless router significantly reduces the QoS. In spite of all these, the researchers still recommended VoIP as an alternative to the traditional circuit switching telephony.

An analysis of the security and challenges of implementing VoIP was conducted by [1]. The review article involved a discussion of the big ideas in VoIP, as well as a sample case study. Also discussed in the article were the advantages and disadvantages of the technology. The author concluded that the quality of VoIP is only as good as the quality of the network link and attributed degraded voice quality to packet delay and loss and their variation.

A fully functional VoIP laboratory using the H.323 protocol with the ability to make calls was implemented by [3]. The research set about obtaining an understanding of the different components and standards of an IP Telephony solution, as well as discussing the various industry standard deployment models for Cisco's IP Telephony. The research achieved a successful implementation of the IP telephony with both H.323 and Media Gateway Control Protocol (MGCP) using the hardware resources at the EECS Networking Laboratory.

The work presented in [4] described the research experiences and results of creating a simulated Ethernet and wireless network scenarios for VoIP in order to determine the suitability of VoIP as a substitute over PSTN. Different LAN, WAN, WLAN, and WWAN scenarios were simulated in OPNET 16 and the result of the simulation was analyzed in terms of the drawbacks of VoIP such as jitter, delay/latency, packet loss, and voice compression. The results were all within the acceptable range for all the scenarios tested and the researchers recommended VoIP as a good substitute to PSTN.

Several simulation experiments by [5] using Network Simulator 2 and some performance parameters (including adding gateways, number of voice traffic flows and the number of hops between source and destinations) were deployed to evaluate the performance of VoIP strategies for hybrid Mobile Ad-hoc Networks (MANETs). Two different cases were evaluated for voice traffics: voice calls between two MANET nodes and voice calls between MANET nodes and external nodes. The key findings of the simulations reveal that adding gateways is not always beneficial. While adding more gateways improve the performance for outgoing traffic flows, however it is not an improvement for intra flows. In addition, adding traffic flows reveal that the network performance is much worse as it causes an increase in the congestion.

A simulation platform to support the planning and design phase of packet switched networks carrying voice traffic was developed by [6]. In order to successfully carry out the research, the author conducted a survey of existing network planning and simulation platforms specifically speech codecs and speech assessment model and created simulation models relevant to the simulation of VoIP traffic; specifically the creation of speech background traffic generation models and creation of simulation models that allow the integration (injection, collection and playout) of real speech data to the simulation. Specifically, the research focused on the creation of a speech background traffic generation models with the purpose of generating traffic that follows the behaviour of a number of speech encoding algorithms.

In addition to the above, this work drew inspirations from similar efforts by some institutions to implement internet telephony in their campus. Specifically, the University of Virginia is in the process of replacing the existing voice communication infrastructure with a new VoIP based system. The institution's VoIP steering committee, in justifying the switch listed a number of functional and technological highlights of the technology and gave an implementation timeline of between January 2014 and December 2015 [7]. Similarly, the University of Texas at Austin recently decided to replace its existing telephone system with a new standard-based enterprise VoIP-based system [8]. The new system has been purchased, and installation of the core was completed in fall 2012. Deployments to buildings across the campus began in spring 2013 and the deployment is expected to be completed in approximately two years thereby completing the transition by 2016 and accelerating features and cost savings for units.

# 3. MATERIALS AND METHODS

The techniques of the Structured Systems Analysis and Design Methodology (SSADM) and Prototyping were adopted in this research. The Analysis phases of the SSADM were adopted in carrying out a detailed study of the present telephony system in our institution while a prototype VoIP system was implemented using Simulation. In addition to the research methodologies, two physical connectivity options were adopted in this research:

i. In the first option, we used the Cisco IP Softphone option. In this case, a JTAOP application running on a computer was the phone used (Figure 1). This helped to reduce the need for separate handsets.

ii. In the second option we connected the IP Phone to the switch and then connected the data device to the switched Ethernet port on the IP Phone (Figure 2). In this way, we were able to use a single port on the switch to provide connectivity to both devices.

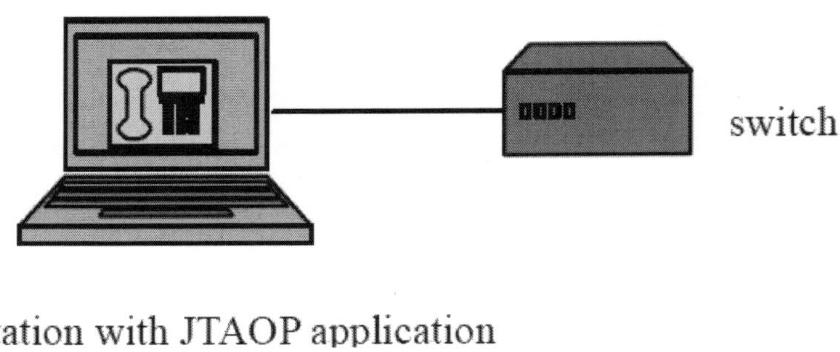


**Figure 1.** Physical connectivity Option1: Cisco IP Softphone option

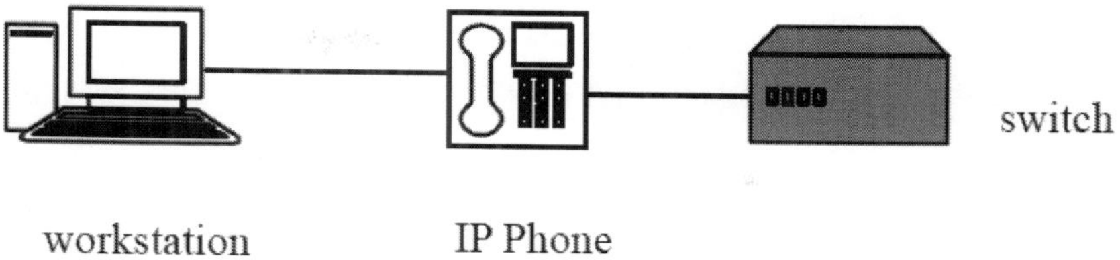


**Figure 2.** Physical connectivity Option2

The materials deployed for this research are:

- Freely available Cisco Packet Tracer Version 6.0.1.0011 was used to run the specification and simulation of the system's behavior. This version provided support for several simulated Application Layer protocols and basic routing with RIP, OSPF, EIGRP and BGP for modeling production networks.
- The following devices were selected during the simulation:
    - Cisco Router 2811
    - Cisco 6500 Layer 3 Switch
    - Cisco 2960 Switch
    - Cisco IP phone 7960 with a large pixel based LCD display, full duplex speaker phone and an internal Cisco Ethernet switch.
- iii. The G.729 voice codec was used for the configuration of this VoIP simulated system as it uses little bandwidth and the quality of voice streaming is high quality

## 3.1. Simulation of the VoIP Design

The Cisco Packet Tracer Simulation of the VoIP Telephony system is given below:

Step 1: Devices were selected from the device database and connected

Step 2: The VLANs and IP networks were created for different user segments and network traffic

Step 3: IP addressed were assigned to the three different endpoints on the campus via the DynamicHost Configuration Protocol. Table 1 below shows the IP Addressing schematic and the respective Virtual LAN (VLAN) used in the design of the VoIP telephony system for both voice and data network.

Step 4: virtual trunking was implemented between the switches and routers and routing was configured between them.

Step 5: The telephony-service was then implemented by configuring the Cisco IP phones and the VoIP dial-peers.

**Table 1.** IP Addresses and Vlan

| Location | | IP Address \|\| VLAN |
|---|---|---|
| (A) Admin Block | Voice | 10.10.10.0/23 \|\| 10 |
| | Data | 10.10.8.0/23 \|\| 8 |
| (B) Library | Voice | 10.10.14.0/23 \|\| 14 |
| | Data | 10.10.15.0/23 \|\| 15 |
| (C) Departments | Voice | 10.10.16.0/23 \|\| 16 |
| | Data | 10.10.17.0/23 \|\| 17 |

# 4. RESULTS AND DISCUSSIONS

## 4.1. Network Design Specification

The simulation of the VoIP system consisted of four network design model; one for each of the three sites listed in Table 1 and the multiple site central model.

- The Admin Model was designed as a star topology with a central switch through which all other connections are made (Figure 3).
- ii. The Library Model was also designed in a star topology but with less computers and IP phones than the administrative block (Figure 4). This is due to the fact that there are less numbers of users in the library site than the administrative block site.
- iii. The Department Model: In order to simplify the simulation, the bus topology was chosen for the department because a prototype VoIP system was designed and simulated for one just one department and this can be re-used for other departments (Figure 5).
- iv. The Multiple Site Central model comprises of IP phones located at remote sites registered with the call manager cluster located at a central site. The star topology is deployed in this mode as other sites are connected to the central location (Figure 6).

## 4.2. Application Design Specification

The overall IP telephony process is represented below using a flowchart (Figure 7).

The first activity is the opening of the data connection, followed by sending a call request. A returned data shows the completion of the call and the connection is closed, otherwise, the call request is re-sent until a data connection is returned.

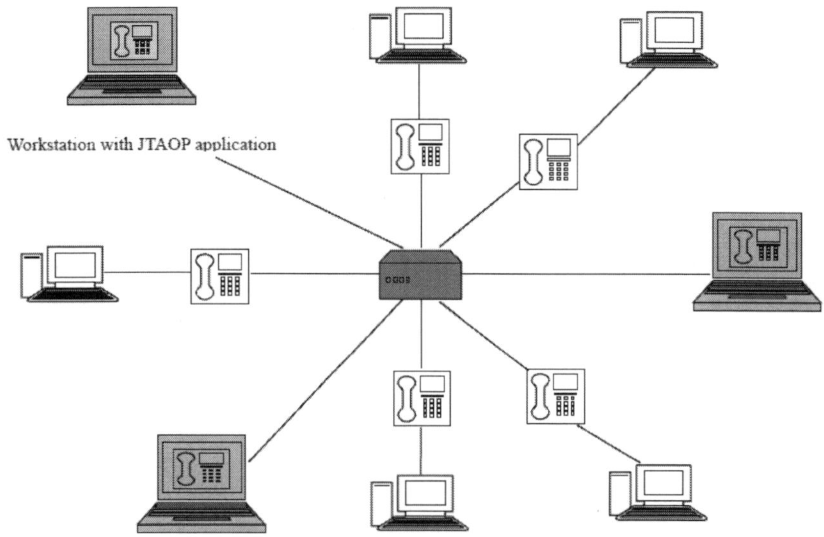


**Figure 3.** Star topology of Admin block

Figure 8 gives the state diagram of the IP telephony system.

## 4.3. Simulation Results and Sample Experimental Test Scenario Visualization

Sample Simulation Results of the VoIP System from Cisco Packet Tracer and phone connection procedure between two clusters are showed below: A visualization of the Cisco Packet Tracer of the simulation of the three sites in the highest hierarchy is showed in figure 9. The IP addresses for voice and data for the three sites are also specified.

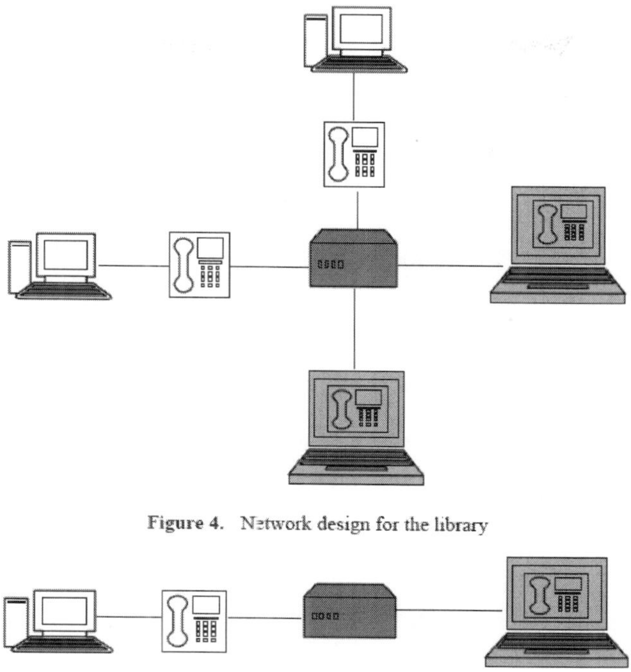


**Figure 4.** Network design for the library

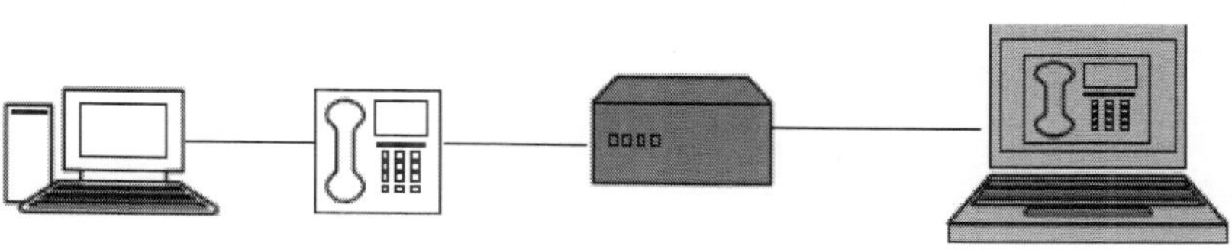

**Figure 5.** Network design for a single department

The connected cluster of devices are simulated and visualized in figures 10 above. Each hub descends from the main router through a main hub in a hierarchical model. The complete configurations and network parameters for each node can be visualized as depicted in Figure 11. Further simulation of the IP handset is presented in figures 12 and 13.

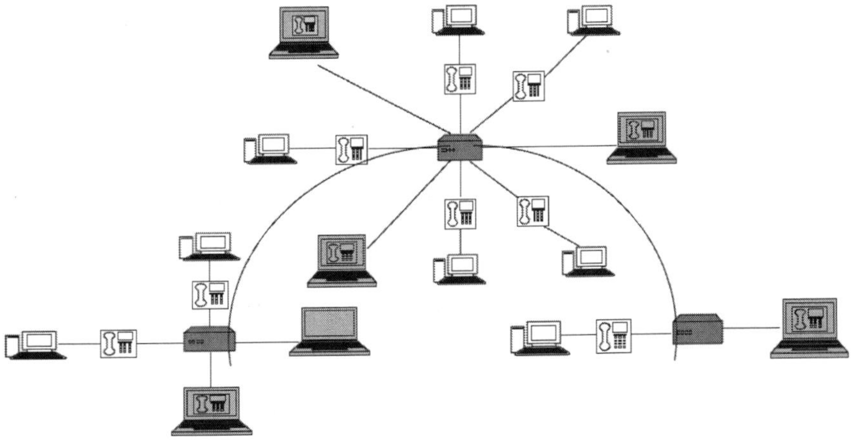

**Figure 6.** Multiple Site model with centralized call processing

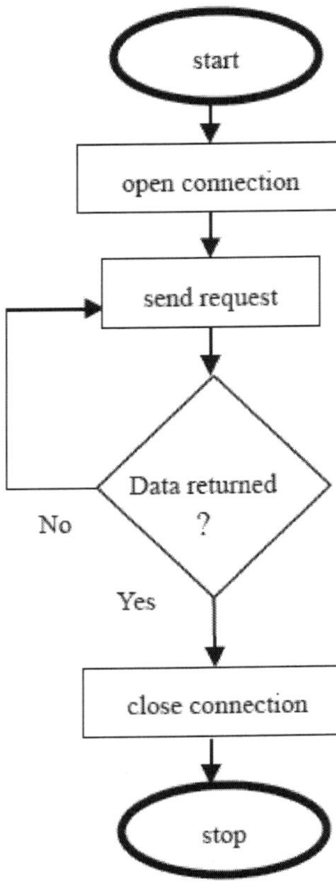


**Figure 7.** Flowchart for the VoIP System

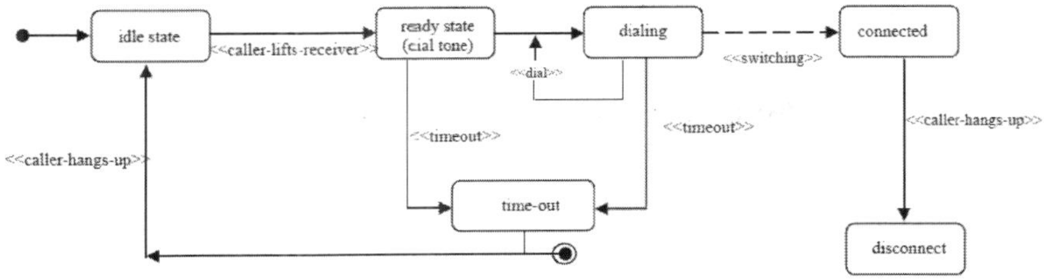


**Figure 8.** State Diagram for the Simulation of the Cisco IP Softphone calling process

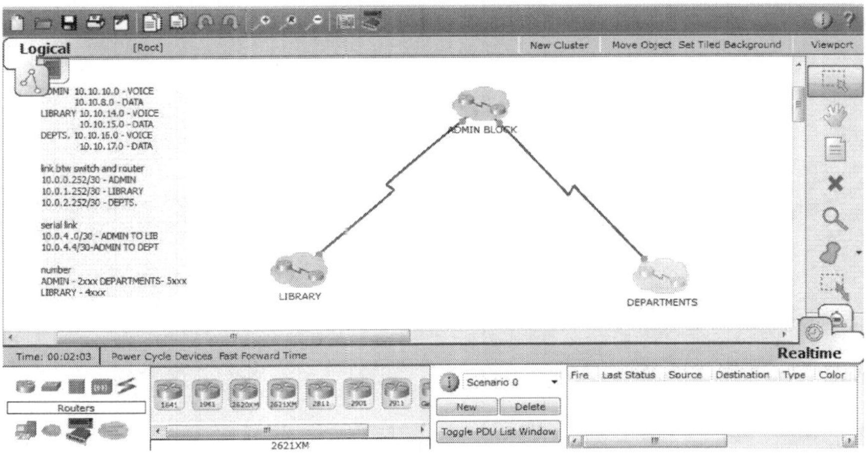


**Figure 9.** Diagram showing simulation of clusters in the VoIP system

In figure 12, extension 4003 dials extension 5001 which then uses SCCP to contact the CUCM to tell the IP phone what to do.

Next, the IP phone contacts the default-gateway to connect it with its destination endpoint (Figure 13).

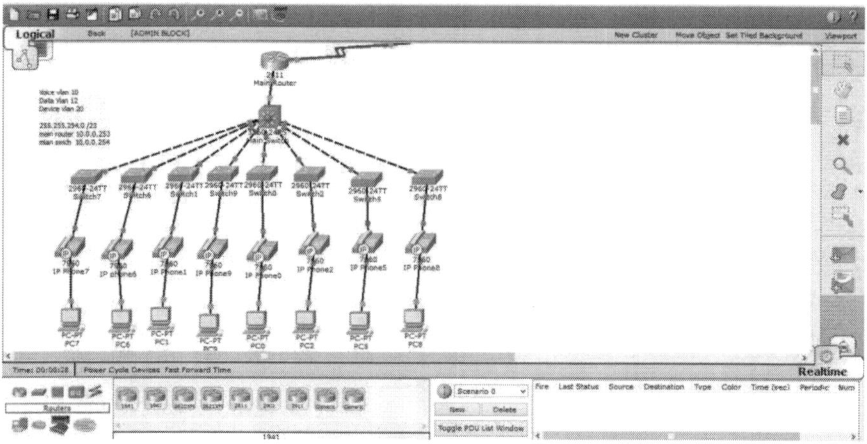


**Figure 10.** Connected Devices in the Admin cluster

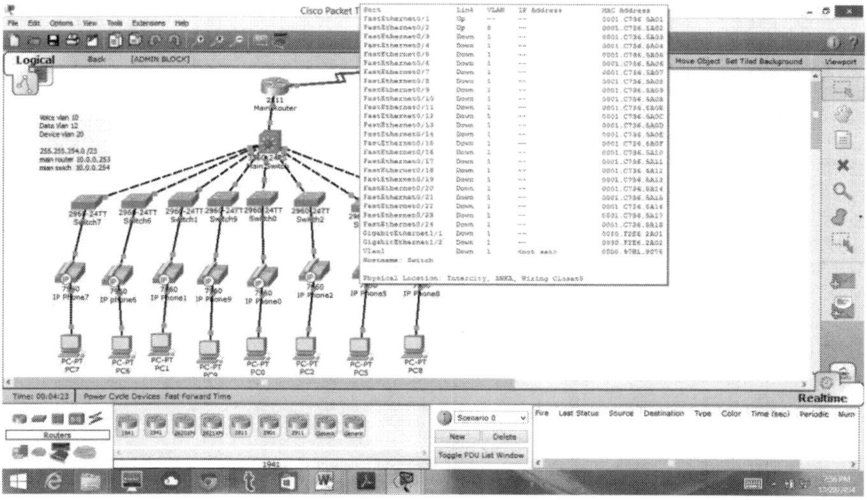


**Figure 11.** Configuration settings for the Admin Switch

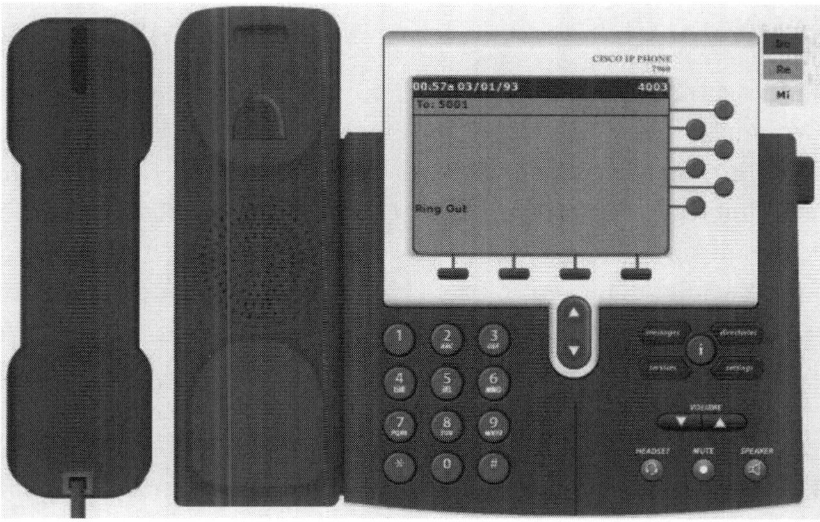


**Figure 12.** Extension 4003 dials extension 5001

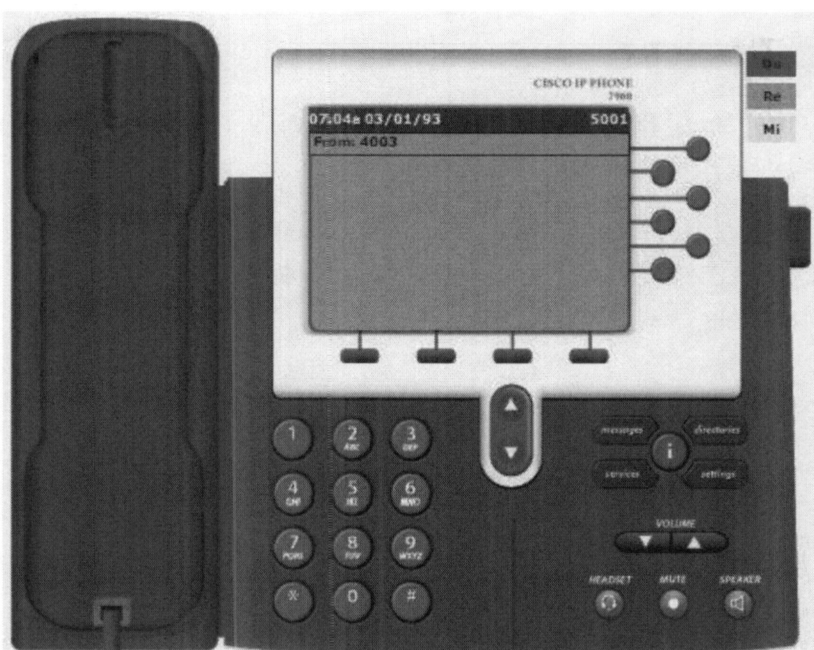


**Figure 13.** Destination (Ext 5001) receives call

# 5. CONCLUSIONS

The researchers in this work had examined the underlining technologies of Voice over Internet Protocols (VoIP) and the possibilities of implementing a campus-wide telephony system using the technology. A prototype design was specified and a simulation was ran using the Cisco Packet Tracer to demonstrate that VoIP can be successfully deployed to provide in a flexible manner, additional data-driven services in campus-wide telephony through a merger of telephone and information technology facilities. While this work had demonstrated the feasibility of leveraging on the affordances of IP-enabled telephones and developing a VOIP-based campus-wide telephony, further research is recommended to test for voice quality and network performance as well as ability for network capacity planning when the system is fully implemented.

# REFERENCES

1. Kumar, A. (2006), An Overview of Voice over Internet Protocol (VoIP). Rivier college online academic journal 2(1).

2. Pan, Y., Chung, J., Zhang, Z. (2012), Analysis of Performance of VoIP Over various scenarios OPNET 14.0. Group 11 Final Report on ENSC 427 Communication Networks, Simon Fraser University Spring 2012 http://www.sfu.ca/~ypa11/Ensc%20427/ 427. html.

3. Kaul, A. (2012), IP Telephony: Review and Implementation. M.Sc. thesis, Department of Electrical Engineering and Computer Science and the Faculty of the Graduate School of the University of Kansas, USA.

4. Dilekci, D., Wang, C., Feng Xu, J. (2013), The Analysis and Simulation of VoIP Group 3 Final Project Report on ENSC 427

Communication Networks, Simon Fraser University. Spring 2013 http://www.sfu.ca/~cjw11/427Project.html.

5. Aguiño, G. I., (2007), Performance of VoIP strategies for hybrid Mobile Ad Hoc Networks. Masters thesis, Department of Computer Science, Karlstad University.

6. Rabassa, A. H. (2010), Simulation Platform for the Planning and Design of Networks Carrying VoIP Traffic Masters of Applied Science in Electrical Engineering Thesis, Faculty of Graduate Studies and Research, Department of Systems and Computer Engineering, Ottawa-Carleton Institute for Electrical and Computer Engineering, Carleton University, Ottawa, Canada.

7. UV (2014), New University Telephone System: UV's Transition to Voice over Internet Protocol (VoIP). The University of Virginia, http://its.virginia.edu/telephone/home.html.

8. UT (2014), Project: Voice over Internet Protocol. University of Texas at Austin, http://www.utexas.edu/its/voip/.

6

# QOS-AWARE COMPOSITE SCHEDULING USING FUZZY PROACTIVE AND REACTIVE CONTROLLERS

**Nabeel Khan, Maria G Martini and Dirk Staehle**

[1]Faculty of Science, Engineering and Computing, Kingston University, Penrhyn Rd, Kingston upon Thames, Surrey KT1 2EE, UK

[2]DOCOMO Euro-Labs, Landsberger Str. 312, 80687 Munich, Germany

## ABSTRACT

We consider in this paper downlink scheduling for different traffic classes at the MAC layer of wireless systems based on orthogonal frequency division multiple access (OFDMA), such as the recent 3rd Generation Partnership Project (3GPP) long-term evolution (LTE)/LTE-A wireless standard. Our goal is to provide via the scheduling decisions quality of service (QoS), but also to guarantee fairness among the different users and traffic classes (including delay-sensitive and best-effort traffic). QoS-aware scheduling strategies, such as modified largest weighted delay first (M-LWDF), exponential (EXP), exponential proportional fair (EXP-PF), and the log-based scheduling rules, prioritize delay-sensitive traffic by considering rules based on the head-of-line (HoL) packet delay and the tolerated packet loss rate, whereas they serve best-effort traffic by considering the classical proportional fair (PF) rule. These scheduling rules do not prevent resource starvation for best-effort traffic. On the other side, if both traffic types are scheduled according to the PF rule, then delay-sensitive flows suffer from delay bound violations. In order to fairly distribute the resources among different service classes according to their QoS requirements and channel conditions, we employ the concept of fuzzy logic in our scheduling framework. By employing the fuzzy logic concept, we serve all the traffic classes with one priority rule. Simulation results show better intra-class and inter-class fairness than state-of-the-art scheduling rules. The proposed scheduling framework enables to appropriately balance delay requirements of traffic, system throughput, and fairness.

# 1 INTRODUCTION

3rd Generation Partnership Project (3GPP) Release 8, and its subsequent modifications, define the long-term evolution (LTE) standard [1] that will take the cellular technology in the 2020s. In wireless communication systems, radio resources are shared by multiple users; hence, medium access control (MAC) layer scheduling becomes extremely important in determining the overall performance of an LTE system. The efficiency of the link level, from the LTE base station (eNodeB) to the mobile terminal, largely depends on the design of the scheduler, whose task is to determine which users should be served and to assign resources.

An efficient scheduler must ensure a good trade-off between efficiency and fairness in the system (according to the service needs of each user) by fully utilizing the available radio resources. MAC layer scheduling strategies can be classified as channel-aware and channel-unaware, where channel aware scheduling algorithms take channel conditions into account and maximize the system throughput. Note, however, that the main target of mobile operators would be the end-user satisfaction, not merely the maximization of system throughput. Scheduling in the LTE standard is more challenging than in earlier standards mainly because earlier standards were based on single carrier systems, where resources were usually divided in terms of time slots or codes among the users, whereas LTE is a multicarrier system where system resources are shared among users in terms of time and frequency sub-bands.

Some approaches solve the problem of resource allocation using optimal solutions, and in other cases, resource allocation and resource assignments are performed in two separate steps; other approaches simply target at adapting schemes originally proposed for time division multiple access (TDMA) to orthogonal frequency division multiple access (OFDMA) systems. Thus, scheduling solutions available in the literature broadly fall into three classes.

1) In [2, 3, 4, 5], resource allocation is modeled as a convex optimization problem. The water-filling algorithm is used to solve the convex optimization problem by considering a continuous objective function. Linear integer programming is also widely used in solving the resource allocation problem by first transforming the nonlinear optimization problem into a linear problem. The main drawback of these strategies is the high computation complexity. Since the transmission time interval (TTI) in LTE is only 1 ms, these algorithms are not feasible from an implementation point of view.

2) In the second class of approaches, such as in [6, 7, 8], scheduling is performed in two steps. The first step consists of resource allocation, which determines the number of resources allocated to each user. The resource allocation step is followed by the resource assignment step, which determines which resources are assigned to each user. This class of scheduling algorithms are specifically designed for delay-sensitive applications and does not provide a priority differentiation between delay-sensitive and best-effort flows.

3) The third approach is the adaptation of TDMA strategies for OFDMA systems. Scheduling rules designed for single carrier systems such as the proportional fair (PF) [9], modified largest weighted delay first (M-LWDF) [10], and exponential proportional fair (EXP-PF) [11] are adapted for an OFDMA system by calculating these rules on each resource. This adaptation is referred to as an OFDMA/TDMA strategy. These scheduling rules are analyzed by [12] for delay-sensitive applications over an LTE system. According to [12], M-LWDF is the best scheduling rule for delay-sensitive applications in terms of fairness and efficiency. A very good survey on these scheduling strategies for LTE is provided in [13]. As each of these scheduling rules are based on the proportional fair rule, the calculation of these scheduling metrics on each physical resource block (PRB) allows the exploitation of multi-user time and frequency diversities. The complexity of the OFDMA/TDMA approach grows linearly with the number of users and resources. Thus, it can be implemented in real time. However, for delay-sensitive traffic, these scheduling rules cannot provide fairness for users with relatively low signal-to-interference noise ratio (SINR) [14].

In this work, we address the following issues of the third class of strategies:

- Intra-class fairness issues for delay-sensitive traffic: scheduling rules for delay-sensitive traffic consider the ratio of instantaneous channel quality and time-averaged throughput (proportional fair rule) along with either the linear [10], logarithmic [15], or exponential [11, 15] function of the head-of-line (HoL) delay [16]. The HoL delay refers to the amount of time packets that reside in the buffer and is also known as the sojourn time. It is important to note that the video is delay-sensitive traffic; hence, packets arriving late are generally not useful at the receiver. Therefore, packets are associated with a predefined HoL delay bound and packets violating the delay bound are dropped from the queue. The utilization of HoL delay and the proportional fair rule in the scheduling decisions are not sufficient to avoid delay bound violation of flows having lower channel quality. Video traffic exhibits highly variable bit rate characteristics, i.e., the instantaneous peak rate is higher than the average rate. Lower channel quality video flows exhibiting peak instantaneous rate have high probability of delay bound violation mainly because of the proportional fair rule in the scheduling decisions. In other words, these scheduling rules achieve higher HoL delay for the packets of flows having higher average rate and lower channel quality. On the other hand, flows having good channel quality and lower average rate are scheduled way before their delay bound. The probability of delay bound violation of the flows exhibiting lower channel quality and higher average rate is very high which results in an unfair system.

- Inter-class fairness issues: in the literature [13], composite scheduling rules serve the best-effort traffic by using the classical proportional fair rule, i.e., ratio of instantaneous channel quality to the time-averaged throughput [9, 17, 18, 19]. They prioritize delay-sensitive traffic by considering either the logarithmic, exponential, or linear function of the HoL delay. However, such composite scheduling strategies result in a higher priority difference between the delay-sensitive and best-effort traffic classes. In other words, the higher the difference among the scheduling priorities of traffic classes, the lower will be the multi-user channel diversity exploitation. In LTE, multi-user channel diversity has more significance since it is a multi-carrier system which allows multi-user diversity exploitation in the time and frequency domain.

By using the concept of fuzzy logic priority [20], we couple the flow's delay urgency (ratio of packet's HoL delay and delay bound) with the time-averaged channel quality. Instead of exploiting the time-averaged throughput and the linear, logarithmic, or exponential function of the HoL delay, we use a fuzzy function of the HoL delay coupled with time-averaged channel quality as introduced in [21]. In

[21], the HoL delay along with the time-averaged channel quality is processed by a fuzzy proactive controller. Further, whenever a flow suffers a delay bound violation, the scheduler reacts to this event and increases the priority of that flow. The delay bound violation input is processed by a fuzzy reactive controller. In this work, we propose a composite scheduling rule for delay-sensitive as well as the best-effort traffic. In the earlier work, the scheduling rule considers only the video traffic. In this work, the scheduling rule and scenarios are extended to handle more than one delay-sensitive traffic types. Furthermore, the main goal of the proposed composite scheduling rule is to balance the probabilities of quality of service (QoS) violation of the delay-sensitive as well as the best-effort traffic types.

A block diagram representing the proposed fuzzy composite scheduling (FCS) is given in Figure 1. The scheduling metric comprises a time-domain priority component based on reactive and proactive controllers and a frequency domain priority based on detailed information on instantaneous channel quality indicator (CQI) feedback per PRB. In order to dynamically adjust the priority level between best-effort and delay-sensitive flows, we utilize a fuzzy-based dynamic resource controller (DRC) (discussed in Section 3.3), as shown in Figure 1. Intra-class fairness (fairness in terms of achieved QoS among the flows within each of the traffic classes) is provided by the fuzzy proactive and reactive controllers whereas inter-class fairness (priority differentiation between the delay-sensitive and best-effort flows) is provided by the DRC. In fuzzy logic, the *output fuzzy set* is defined as the range of all possible output values that can be assigned to a fuzzy controller. The output of the controller lies within the output fuzzy set. The larger the output fuzzy set, the higher the priority of the controller. In the proposed scheduling framework, each traffic class has its own output fuzzy set. We assign a fixed output fuzzy set to the delay-sensitive traffic class and dynamically adjust the output fuzzy set of the best-effort traffic. The output fuzzy set of the best-effort traffic class is set by the DRC based on the latency (packet's HoL delay) and QoS violation of the delay-sensitive flows as shown in the figure. The higher the latency and QoS violations of the delay-sensitive flows, the lower the output fuzzy set of the best-effort traffic. The final priority on each PRB is a function of the time and frequency domain priority metrics as shown in Figure 1.

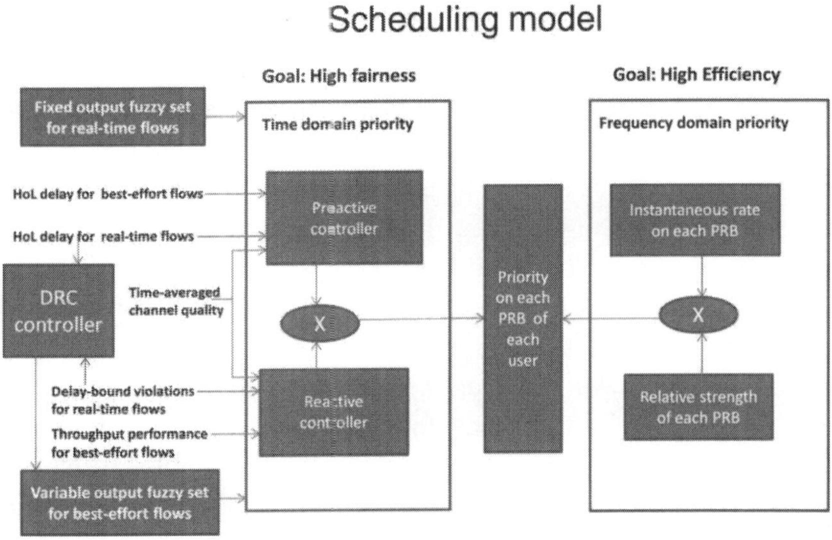


**Figure 1.** Time and frequency domain models of the FCS scheduling framework.

The remainder of this paper is organized as follows. Section 2 presents the considered system model and the problem statement. Section 3 presents the details of our fuzzy logic-based scheduling strategy. Section 4 presents the performance evaluation of the proposed approach. In particular, the solutions considered as benchmark for the assessment of our scheduling algorithm are presented in Section 4.1, whereas the simulation setup is presented in Section 4.2; results are presented and discussed in Section 4.3. Conclusions are drawn in Section 5.

## 2 SYSTEM MODEL AND PROBLEM STATEMENT

We consider a multiuser downlink single input single output (SISO) LTE/LTE-A system. The single-cell scenario comprises an eNodeB MAC scheduler responsible for allocating PRBs to different users in the cell. Each user *i* is assigned a buffer at the eNodeB, and packets of different traffic classes are streamed into the buffer of the eNodeB. For delay-sensitive traffic, we consider video and VoIP traffic (the scheduling framework can accommodate all LTE service classes), whereas for best-effort traffic, we consider constant bit rate (CBR) traffic. The packets of each traffic class entering the buffer are time stamped by the scheduler. Packets of delay-sensitive traffic are dropped from the buffer if the HoL packet delay is longer than the target HoL delay bound. The main QoS parameters for video and VoIP flows are the HoL packet delay and the packet loss rate (PLR), whereas throughput is the important QoS parameter for the flows of best-effort traffic. We consider the HoL delay for best-effort traffic, and we assign a target delay for the flows of this traffic class. However, since we can assume best-effort traffic is delay tolerant, therefore, packets violating the target HoL delay are not dropped from the buffer. We use a CQI feedback mechanism, where each user feedbacks information about the channel quality on each PRB. Due to the adoption of adaptive modulation and coding (AMC) in LTE, each CQI value corresponds to a specific value of spectral efficiency for each PRB.

At scheduling epoch *n*, we define the normalized time-averaged wideband spectral efficiency, $\overline{\chi}_i^{(n)}$, of user *i* over the moving average window of size $n_c$ as:

$$\overline{\chi}_i^{(n)} = \frac{1}{\chi_{\max}} \left[ \frac{1}{n_c} \sum_{k=n-n_c}^{n} \chi_i^{(k)} \right] \tag{1}$$

where

$$\chi_i^{(n)} = \frac{1}{M_{\text{PRB}}} \sum_{\varphi=1}^{M_{\text{PRB}}} \chi_{i,\varphi}^{(n)} \tag{2}$$

is the average PRB spectral efficiency of user *i* at scheduling instant *n* and $\chi_{i,\varphi}^{(n)}$ is the instantaneous subband spectral efficiency of user *i* at PRB $\varphi$. $\chi_{\max}$ is a constant, i.e., the spectral efficiency (5.5547 bits/s/Hz) corresponding to the maximum CQI feedback, and $M_{\text{PRB}}$ is the number of PRBs available for allocation at each scheduling epoch.

Given the available information about:

- the HoL packet delay for each flow $H_i^{(n)}$,
- the channel quality of each flow on each PRB, hence the resulting spectral efficiency $\chi_{i,\varphi}^{(n)}$,
- the tolerated delay bound $H_{\max}$,
- the QoS performance of the delay-sensitive flows in terms of packet loss ratio, $\text{plr}_i^{(n)}$ and of the best-effort flows in terms of time-averaged throughput $R_{i,\text{ave}}^{(n)}$,

the scheduling problem is defined as: *How to allocate to the different users the M $_{PRB}$ PRBs in each scheduling interval in order to fulfill the QoS requirements of each of the flows from different traffic classes so that a good trade-off between fairness and efficiency is achieved.*

In order to mathematically formulate the problem, let us define the following parameters:

$R_i^{(n)}$: Throughput achieved by flow *i* at scheduling instant *n*.

*I*: Total number of flows in the system. It is the sum of delay-sensitive $I_{\text{delay-sensitive}}$ and best-effort $I_{\text{best-effort}}$ flows.

$\text{plr}_i^{(n)}$: The packet loss ratio of flow *i* at scheduling instant *n* calculated over the moving average transmission window $t_w$:

$$\text{plr}_i^{(n)} = \frac{\sum_{m=n-t_w}^{n} P_{\text{drop}_i}^{(m)}}{\sum_{m=n-t_w}^{n} \left(P_{\text{transmit}_i}^{(m)} + P_{\text{drop}_i}^{(m)}\right)} \quad (3)$$

where

$P_{\text{transmit}_i}^{(m)}$: Number of transmitted packets of flow *i* over the moving average transmission window $t_w$.

$P_{\text{drop}_i}^{(m)}$: Number of dropped packets of flow *i* over the moving average transmission window $t_w$.

The main goal of the scheduler is to maximize the system through put $R_{t_w}^{(n)}$, subject to the QoS constraints of the delay-sensitive flows, over the moving average transmission window $t_w$:

$$R_{t_w}^{(n)} = \max\left(\sum_{i=1}^{I} \sum_{m=n-t_w}^{n} R_i^{(m)}\right) \quad (4)$$

subject to

$$\frac{1}{I_{\text{delay-sensitive}}} \sum_{i=1}^{I_{\text{delay-sensitive}}} \mathbb{1}\left(\text{plr}_i^{(n)} \leq \text{plr}_{\text{thr}}\right) = 1 \quad (5)$$

where

$\mathbb{1}\left(\text{plr}_i^{(n)} \leq \text{plr}_{\text{thr}}\right)$ is an indicator function equal to 1 if its argument is true, i.e., when the packet loss rate of flow *i* is lower or equal than the threshold value $\text{plr}_{\text{thr}}$. If the packet loss rate exceeds the threshold, then the indicator function is 0. It is important to note that fairness for delay-sensitive traffic is guaranteed when the PLR over a short moving average window [22], for instance one second, is below the prescribed threshold for each of the delay-sensitive flows in the system. As noted in [23], when the scheduler achieves short-term fairness, then the long-term fairness is guaranteed.

The optimal solution of the above problem is not possible without restrictive assumptions on the arrival process of the traffic and changes in channel quality. Therefore, we propose a novel scheduling framework relying on fuzzy logic. Fuzzy logic is ideally suited for problems where a definite mathematical solution is unavailable. The information about the changes in the radio channel and the traffic rate of each user is uncertain. Fuzzy logic can deal with such situations because of its capability to make approximate reasoning. In our proposed scheduling strategy, each PRB is assigned to the user maximizing a defined metric. Our proposed metric is composed of a PRB-independent part and a PRB-specific part. The PRB-independent part calculated for a user describes the 'urgency' of an assignment as time-domain priority, whereas the PRB-specific part describes the instantaneous channel quality of the PRB and its relative quality versus other PRBs.

## 3 FUZZY COMPOSITE SCHEDULING FRAMEWORK

The FCS framework consists of fuzzy proactive, reactive, and DRC controllers. It is important to note that the designs of the proactive and reactive controllers are the same. The proactive controller processes the HoL delay whereas the reactive controller processes the QoS violation. In the following, we present a detailed design of the three fuzzy controllers:

### 3.1 Proactive controller

The goal of the proactive controller is to avoid delay bound violations. In order to consider the delay urgency in a dynamic wireless environment, we propose a novel concept of utilizing time-averaged channel quality over a small moving window by using the average wideband spectral efficiency $\overline{\chi}_i^{(n)}$ associated to the CQI feedback, defined in Equation 1. The proactive controller processes two inputs. One of these is the HoL packet delay $H_i^{(n)}$ normalized to the maximum tolerated HoL delay $H_{max}$ of each traffic class. The normalized HoL delay input, $H_i'$, for the proactive controller is:

$$H_i' = \begin{cases} \frac{H_{i,\mathrm{VoIP}}^{(n)}}{H_{i,\max}}, & \text{for VoIP traffic class} \\ \frac{H_{i,\mathrm{video}}^{(n)}}{H_{i,\max}}, & \text{for Video traffic class} \\ \frac{H_{i,\mathrm{best-effort}}^{(n)}}{H_{i,\max}}, & \text{for best} - \text{effort traffic class} \end{cases} \quad (6)$$

The goal of the controller is to be proactive for any possible delay violations; hence, the second input is designed as the weighted sum of the normalized delay and the normalized average channel quality. It is mathematically defined as:

$$\chi_{H'} = 0.5(1 - H_i') + 0.5(\overline{\chi}_i^{(n)}) \quad (7)$$

The rationale behind the weighted sum Equation 7 is discussed in Section 3.1.1.

In fuzzy logic, the input membership function represents the magnitude of the inputs which are mapped to the output membership function through a set of rules [20]. The membership functions can be linear, exponential, bell shaped, or any other shape according to the system requirements. According to [12], the M-LWDF scheduling rule, linear function of the HoL packet delay, outperforms the EXP-PF scheduling rule which is an exponential function of HoL packet delay. Therefore, we select linear membership functions for the proactive and reactive controllers. The graphical representation of the input and output membership functions is shown in Figure 2a and 2b, respectively. The same input membership functions are used for both the inputs$(H_i'$ and $\chi_{H'})$. It is important to note that users with better channel quality result in a higher frequency domain priority on each PRB $\varphi$, as there will be a higher number of PRBs with better channel quality. Therefore, the time domain priority should be higher for users with higher normalized HoL packet delay and lower normalized channel quality.

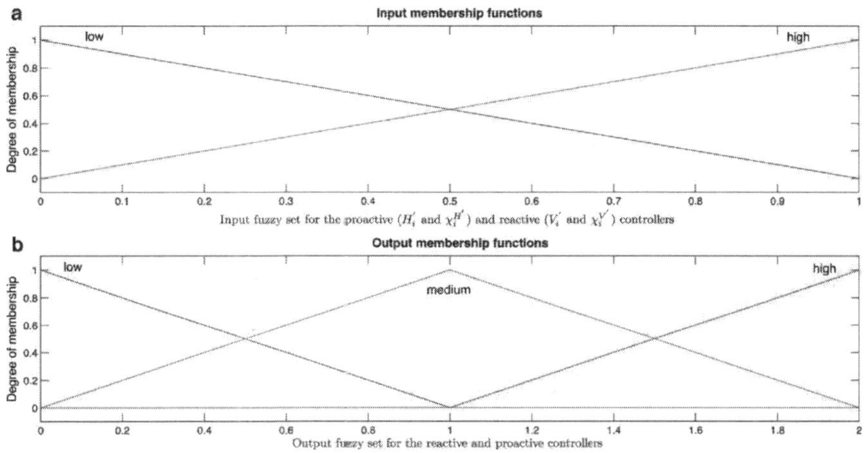


**Figure 2. Input and output membership functions. (a)** Input membership functions low and high of the proactive and reactive controllers. **(b)** Output membership functions low, medium, and high of the proactive and reactive controllers.

Now, we will utilize the flexibility of fuzzy logic by mapping the input membership functions to the output memberships functions through a set of rules. Let $\mu_p$ be the output of the proactive controller (defuzzified proactive priority value), the fuzzy rules for the proactive controller are as follows:

1) If H'i is low AND χH'i is low THEN $\mu_p$ is medium

2) If H'i is low AND χH'i is high THEN $\mu_p$ is low

3) If H'i is high AND χH'i is low THEN $\mu_p$ is high

4) If H'i is high AND χH'i is high THEN $\mu_p$ is medium

where *low, medium*, and *high* are the output membership functions as shown in Figure 2 and $\mu_p$ is the crisp output which along with the reactive controller output quantifies the time domain priority of each user. The main motivation of using the low, medium, and high output membership functions is to prioritize flows suffering from lower channel quality and higher HoL delay. The priority of the users with relatively good channel quality increases from low to medium as the HoL delay increases. On the other hand, the priority of users with lower channel quality increases from medium to high. Therefore, fairness is incorporated in the scheduling decisions through the output membership functions and rules of the fuzzy controllers. The main goal of the frequency domain priority is to improve the system efficiency whereas the time domain priority provides fairness through fuzzy proactive and reactive controllers.

The output fuzzy set of the membership functions, shown in Figure 2, determines the traffic priority of each traffic class. It is important to note that $\mu_p$ lies within the output fuzzy set. The proactive priority, $\mu_p$, as a function of the inputs H′i and χH′i is shown in Figure 3.

The steps involved in producing a crisp output in the fuzzy logic system are described below.

1) Fuzzification. This is the process of converting fuzzy input values into a degree of membership for each linguistic term. Each linguistic term, high, medium, and low, characterizes a membership function. For instance, the proactive controller inputs, H′i and χH′i, as shown in Figure 4, are fuzzified by the input membership functions low and high. In the figure, the four rows are the four rules of the proactive controller. Rule one comprises only low membership function, therefore input H′i and χH′i are fuzzified by the low membership function as shown in the figure.

2) Fuzzy inference. This is the core process of the fuzzy logic system, comprising a mapping from a given input to an output

using the membership functions and logical operators in the *if-then-else* rules. According to Figure 4, the AND logical operation is performed, according to which the minimum of the two fuzzified inputs is mapped to the output membership function. The result of the fuzzy inference process is the degree of the output membership functions fulfilled by the logical operations between the fuzzified inputs. The result is the truncated output membership functions as shown in the third column of Figure 4.

3) 'Defuzzification' and production of the final 'crisp' output. The crisp proactive priority output $\mu_p$ produced is shown in Figure 4. The output of each rule is combined to give the final fuzzy set, as shown in the fifth row and third column in Figure 4. The defuzzification process is simply the centroid calculation on the final fuzzy set as shown in Figure 4.

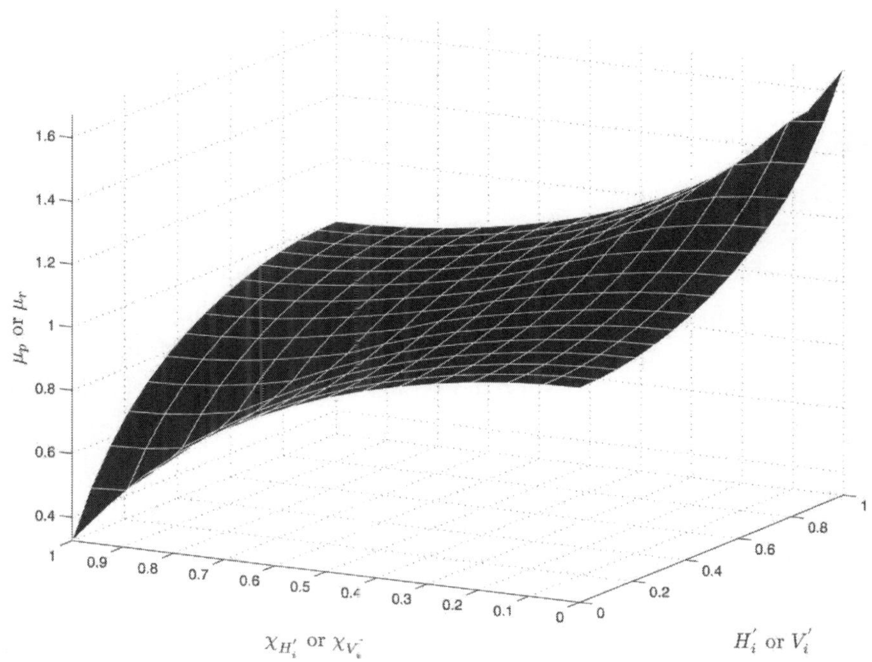


**Figure 3.** Proactive controller output, $\mu_p$, w.r.t the inputs.

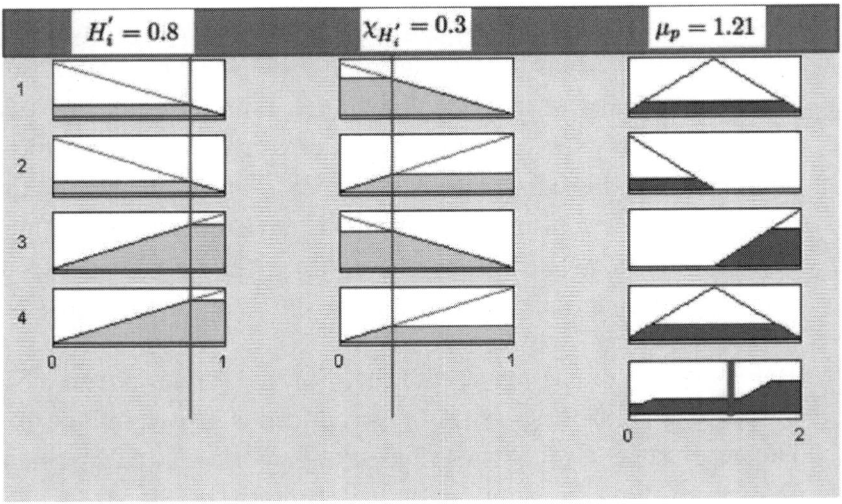


**Figure 4.** Fuzzy rules of the proactive controller.

### 3.1.1 Rationale

The inputs H′i and χH′i are a function of the HoL delay; hence, the system is made more proactive for any possible delay violations. The second input, weighted sum of the normalized HoL delay and time-averaged channel quality, enhances the system fairness. For instance, consider two users - user 1 and 2 - having normalized average channel quality of 0.8 and 0.6, respectively, and normalized HoL packet delay of 0.4 and 0.8, respectively. If we select the second input as a simple function of the average channel quality, i.e., $\chi_{H'_i} = \overline{\chi}_i^{(n)}$, then the output of the proactive controller, $\mu_p$ (Figure 3), for user 1 and 2 is 0.844 (H′i=0.4, χH′i=0.8) and 1.05 (H′i=0.8, χ H′i=0.6), respectively. The difference in the proactive priority of the two users is 1.05-0.844=0.206. On the other hand, if the weighted sum Eq. 7 is used, then the output for user 1 and 2 is 0.872 (H′i=0.4, χH′i=0.6) and 1.14 (H′i=0.8, χH′i=0.7), with a difference in proactive priority of 0.268. A higher priority with weighted sum equation quantifies the urgency in the service needs of user 2 having relatively higher packet delay and lower channel quality. Therefore, the system is more sensitive to the HoL delay. If the instantaneous channel quality of the user improves, the system exploits it. For instance, consider Figure 5 where the channel quality increases at the *current scheduling instant*; the result is a higher time domain priority, quantifying lower time-averaged channel quality over a window of size $n_c$ epochs and higher HoL packet delay. Because of the increase in the channel quality at the current scheduling instant, the frequency domain priority (function of current instantaneous channel quality) also increases with PRBs having better channel quality. Therefore, the weighted sum of the normalized HoL delay and the time-averaged channel quality with weights equal to 0.5 makes the system opportunistic (exploiting instantaneous channel improvements) and delay aware.

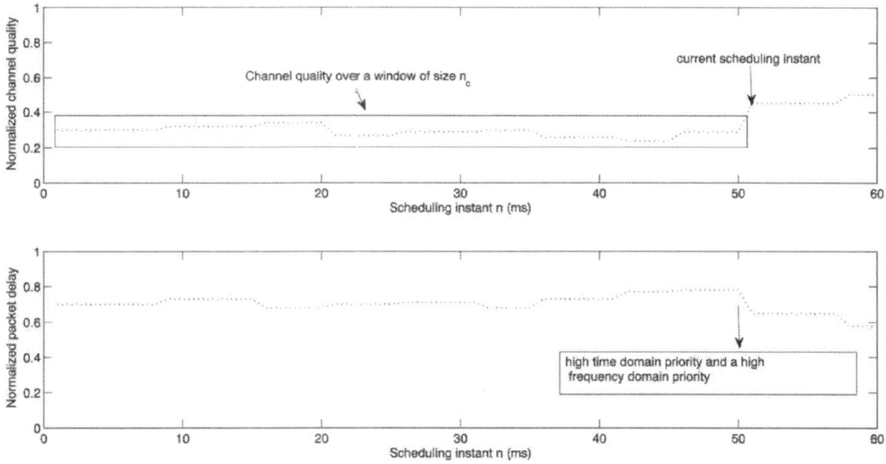


**Figure 5.** Rationale of the proactive controller design.

## 3.2 Reactive controller

Delay-sensitive applications can tolerate packet losses if they are below a given threshold. To provide fairness in multimedia traffic, packet losses should be kept below a given threshold for all users. The goal of the reactive controller is to distribute the packet losses proportionally equal across all the users. In order to define the inputs of the reactive controller, we utilize the packet loss ratio, $\mathrm{plr}_i^{(n)}$, of user $i$ given in Equation 3. The packet loss ratio can easily be calculated by using the number of dropped and transmitted packets over a small transmission window. The design of the reactive controller is similar to the proactive controller except that the fuzzy inputs are based on the packet loss rate over a moving average transmission window. We consider a window size of 1 s in the simulation study. The amount of QoS violation in terms of packet loss ratio and tolerated packet loss threshold, $\mathrm{plr}_{\mathrm{thr}}$, of user $i$ at scheduling instant $n$ is:

$$V_{i,\text{delay-sensitive}}^{(n)} = \frac{\mathrm{plr}_i^{(n)}}{\mathrm{plr}_{i,\mathrm{thr}}}. \tag{8}$$

The QoS parameter for the delay-sensitive traffic is the packet loss ratio, whereas for the best-effort flows, the QoS performance parameter is the ratio of minimum rate required to the achieved time-averaged throughput.

$$V_{i,\text{best-effort}}^{(n)} = \frac{R_{\min}}{R_{i,\mathrm{ave}}^{(n)}} \tag{9}$$

where $R_{i,\mathrm{ave}}^{(n)}$ is the time-averaged throughput and $R_{\min}$ is the minimum rate requirement. The QoS violation input, $V_i'$, for the reactive controller is:

$$V_i' = \begin{cases} \dfrac{V_{i,\mathrm{VoIP}}^{(n)}}{V_{j,\max}^{(n)}}, & \text{for VoIP traffic class} \\ \dfrac{V_{i,\mathrm{Video}}^{(n)}}{V_{j,\max}^{(n)}}, & \text{for Video traffic class} \\ \dfrac{V_{i,\text{best-effort}}^{(n)}}{V_{j,\max}^{(n)}}, & \text{for best-effort traffic class} \end{cases} \tag{10}$$

It is a requirement of the fuzzy logic system that the inputs of the fuzzy controller should lie within the input fuzzy set, i.e., in between 0 and 1. Therefore, we normalize the input with respect to the flow having the maximum QoS violation, $V_{j,\max}^{(n)}$.

The second input, $\chi_{V_i'}$, of the reactive controller is designed as the weighted sum of the normalized QoS violations and the normalized average channel quality. Mathematically, it is defined as:

$$\chi_{V_i'} = 0.5(1 - V_i') + 0.5(\overline{\chi}_i^{(n)}). \tag{11}$$

### 3.2.1 Rationale

The rationale behind the design of the reactive controller is the same as that of the proactive controller discussed in Section 3.1.1. The weighted sum of the normalized QoS violations and the time-averaged channel quality with weights equal to 0.5 makes the system opportunistic (exploiting instantaneous channel improvements) and QoS aware as discussed in Section 3.1.1. The input and output membership functions and the output fuzzy set is the same as that of the proactive controller. It is important to note that we could have used all the inputs, i.e., the HoL packet delay, the QoS violations, and the time-averaged channel quality, and design a fuzzy priority scheme by defining a set of rules for these three inputs. However, this increases the complexity of the system because, with three inputs, eight rules and more than three output membership functions are required. A fuzzy logic system with two inputs is simpler in terms of implementation and processing. Therefore, by using the same rules and membership functions, the same fuzzy module is called for proactive ($H_i'$ and $\chi_{H_i'}$) and reactive ($V_i'$ and $\chi_{V_i'}$) inputs.

## 3.3 Dynamic resource controller

The best-effort traffic class is considered as the lowest priority class. Scheduling rules designed for delay-sensitive traffic, such as in [24, 25] (see the time utility functions of different traffic classes), give low scheduling priority to the best-effort flows. High priority differentiation between the delay-sensitive and best-effort flows causes resource starvation for the best-effort flows [26, 27]. In FCS scheduling framework, inter-class traffic priority differentiation is provided by output fuzzy set. The output fuzzy set represents the range of all possible output values that can be assigned to the proactive and reactive controllers. The larger the output fuzzy set, the higher the priority of the controller. In order to dynamically prioritize flows belonging to best-effort traffic class, we adapt the output fuzzy set of the best-effort flows according to the QoS performance of the delay-sensitive flows. The output fuzzy set of the delay-sensitive traffic class is fixed; thus, the amount of resource allocations between the delay-sensitive and best-effort flows is adaptable and controlled by the maximum limit of the output fuzzy set,$\mu_{max}$, as given in Equation 12.

$$\mu_{r_{\text{best-effort}}} = \mu_{p_{\text{best-effort}}} \in \{0, \mu_{\max}\} \tag{12}$$

where μrbest−effort and μpbest−effort are the defuzzified outputs of the reactive and proactive controllers, respectively. As discussed in Sections 3.1 and 3.2, the design of both the controllers and the corresponding output fuzzy sets are the same. Flows from each traffic class utilize the same time domain priority by using the reactive and proactive controllers. The average delay and packet loss rate performance of the delay-sensitive flows are used to determine the maximum limit of the output fuzzy set for the best-effort traffic. Mathematically, the average QoS parameters of the delay-sensitive flows are as follows:

$$\overline{H_i'} = \frac{1}{I_{\text{delay-sensitive}}} \sum_{i=1}^{I_{\text{delay-sensitive}}} H'_{i,\text{delay-sensitive}} \tag{13}$$

$$\overline{V_i'} = \frac{1}{I_{\text{delay-sensitive}}} \sum_{i=1}^{I_{\text{delay-sensitive}}} V'_{i,\text{delay-sensitive}} \tag{14}$$

where $I_{\text{delay-sensitive}}$ is the number of delay-sensitive users, $\overline{H_i'}$ is the average normalized delay, and $\overline{V_i'}$ is the average QoS violations of all the delay-sensitive users. The input and output membership functions of the DRC are shown in Figure 6a and 6b, respectively. The maximum limit, $\mu_{max}$ is set according to the following fuzzy rules:

1) If $\overline{H_i'}$ is low AND $\overline{V_i'}$ is low THEN $\mu_{max}$ is high

2) If $\overline{H_i'}$ is high AND $\overline{V_i'}$ is low THEN $\mu_{max}$ is low

3) If $\overline{H_i'}$ is high AND $\overline{V_i'}$ is high THEN $\mu_{max}$ is low

4) If $\overline{H_i'}$ is low AND $\overline{V_i'}$ is high THEN $\mu_{max}$ is medium

The input degree of membership is determined by the trapezoidal input membership functions. A lower average packet delay and loss rate causes rule 1 to have a higher degree of membership. Therefore, $\mu_{max}$ is maximum as given by the centroid of the highest area triangle membership function as shown in Figure 6. On the other hand,$\mu_{max}$ is set to minimum when a higher average HoL delay and packet loss rate causes the smallest area triangle to be defuzzified through rule 2 and rule 3. If the normalized average delay is lower and average PLR is higher than the medium area, triangle is defuzzified as given in rule 4.

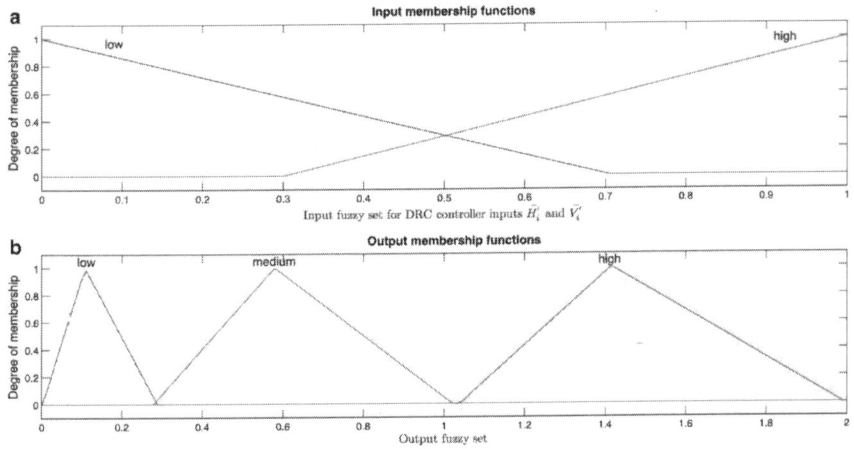


Figure 6. **Input and output membership functions of the DRC. (a)** Input membership functions low and high of the DRC.**(b)** Output membership functions low, medium, and high of the DRC.

### 3.3.1 Rationale

The main rationale of utilizing DRC is to serve the following three goals:

- Utilization of delay tolerant nature of the best-effort traffic: according to the policy guidelines of the QoS architecture in the 3GPP standard, the resource allocation probability of the best-effort traffic class should be minimum in situations where the network becomes congested with delay-sensitive traffic. When the traffic load reaches the network capacity, the increase in average packet's latency of the delay-sensitive traffic decreases the maximum limit of the output fuzzy set for the best-effort flows as shown in Figure 7. Since best-effort traffic is delay tolerant, the decreased maximum limit of the output fuzzy set ensures delay-sensitive traffic gets priority over best-effort traffic.

- Channel diversity exploitation: the main goal of the scheduler is to maximize the system throughput subject to maintaining the deadline violations below the prescribed threshold (Equation 5). At lower normalized average packet latency, the priority difference between the delay-sensitive and best-effort flows is minimal. Hence, flows from different traffic classes are scheduled based on their QoS performance and channel quality. Utilization of same output fuzzy set for the DRC, proactive, and reactive controllers: the prioritization of the delay-sensitive flows w.r.t the best-effort traffic can be achieved by using the same output fuzzy set for the proactive, reactive, and DRC controllers. When the output fuzzy set of these controllers are same, then the increase in latency of the delay-sensitive flows causes a reduction in the output fuzzy set of the best-effort traffic as shown in Figure 7. When the network becomes heavily congested, then delay bound violations occur for the delay-sensitive flows. The delay bound violation further reduces the output fuzzy set of the best-effort traffic as shown in Figure 7. Thus, decreasing the resource allocation probability of the best-effort traffic.

## 3.4 Time domain priority

The proactive controller output, $\mu_p$, and the reactive controller output, $\mu_r$, define the time domain priority of the scheduling rule. Let $\mu_i^{(n)}$ be the final time domain priority which is the product of the output of both the controllers given as:

$$\mu_i^{(n)} = (\mu_p \mu_r)^{\alpha_t} \tag{15}$$

where $\alpha_t$ is the time domain fairness parameter which enables the operator of the system to tune the fairness level. The higher the value of $\alpha_t$, the higher will be the time domain priority of users suffering from relatively poor channel quality, higher HoL delay, and higher QoS violations.

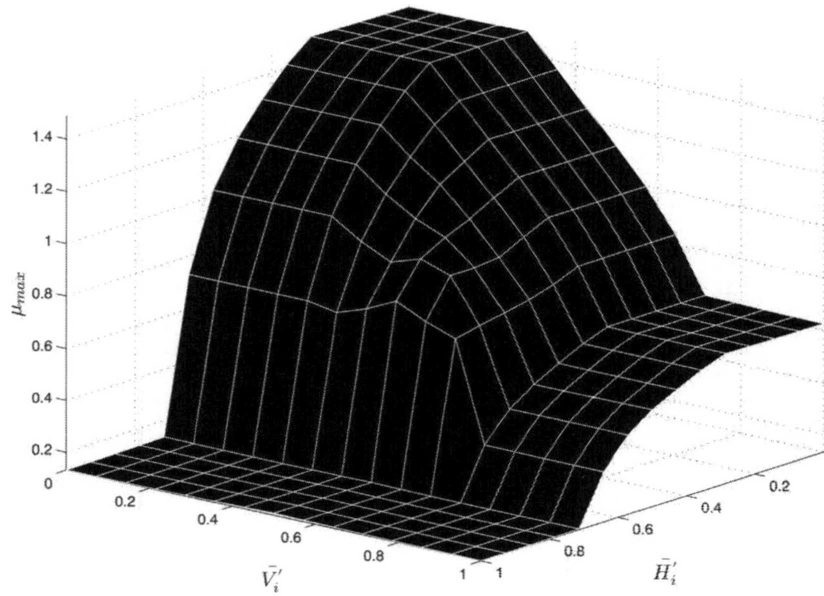


**Figure 7.** DRC's response to the normalized average delay and average PLR.

## 3.5 Frequency domain priority

The time domain priority, by utilizing past and current CQI feedbacks, considers the channel quality over a small window. The goal of the time domain priority is to control the fairness among the users. On the other hand, the goal of the frequency domain priority is to improve the system efficiency by considering only the current CQI feedback. Due to multipath propagation and interference from the neighboring users, there is a variable amount of fading on the PRBs of each user. Efficiency as well as fairness can be enhanced if this information is utilized. By employing the CQI feedbacks on each of the PRBs, information on the interference and multipath propagation can be obtained [28, 29].

Hence, we adopt a parameter called relative strength of user $i$ on PRB $\varphi$ which is given as:

$$\theta_{i,\varphi}^{(n)} = \frac{\chi_{i,\varphi}^{(n)}}{\chi_i^{(n)}} \tag{16}$$

where $\varphi^{\theta_{i,\varphi}^{(n)}}$ gives information on the variable amount of fading on the PRBs of each user. If a user is experiencing a high interference on some of the PRBs, this factor assigns a lower weight to such PRBs. On the other hand, the PRBs with the best channel quality for a user will be assigned a higher weight thus exploiting the independent multi-user frequency selective fading. The frequency domain priority, $\Gamma_{i,\varphi}^{(n)}$, of user $i$ on PRB $\varphi$ is the product of channel quality and relative strength:

$$\Gamma_{i,\varphi}^{(n)} = \chi_{i,\varphi}^{(n)} \theta_{i,\varphi}^{(n)}. \tag{17}$$

Replacing in (17), the expression of $\theta_{i,\varphi}^{(n)}$ given in (16), we get

$$\Gamma_{i,\varphi}^{(n)} = \frac{[\chi_{i,\varphi}^{(n)}]^2}{\chi_i^{(n)}}. \tag{18}$$

## 3.6 Final scheduling priority metric

It has been shown in [15] that a good trade-off between fairness and efficiency can be achieved by defining a priority function which is the product of the logarithmic function of the time-domain priority and a linear function of the instantaneous rate on each PRB. The time-domain priority used in the LOG rule in [15] is a function of the HoL packet delay. We use a time-domain priority which is derived from fuzzy logic and is a function of the user's HoL packet delay, time-averaged channel quality and packet loss rate. The final priority, $\mathrm{PRF}_{i,\varphi}^{(n)}$, of user $i$ on PRB $\varphi$ is a function of the logarithm of the time domain priority and it varies linearly with the frequency domain priority as given below:

$$\mathrm{PRF}_{i,\varphi}^{(n)} = \log(1+\mu_i^{(n)})\,\Gamma_{i,\varphi}^{(n)}. \tag{19}$$

User $i^*$ is allocated a PRB $\varphi$ satisfying the following rule:

$$i^* = \operatorname{argmax}\left(\mathrm{PRF}_{i,\varphi}^{(n)}\right). \tag{20}$$

It is important to note that state-of-the-art scheduling rules serve best-effort flows with the classical delay-insensitive PF rule and prioritize the delay-sensitive traffic by considering the HoL packet delay. We use the same priority equation given in 19 for all the traffic classes; dynamic prioritization between the delay-sensitive and best-effort classes is achieved by using the DRC. More details on the prioritization of different traffic classes is given in the following sections.

# 4 PERFORMANCE EVALUATION

## 4.1 Benchmark scheduling rules

In order to assess the performance of the proposed FCS scheduling rule, we compare it the with the state-of-the-art strategies shown in Table 1.

**Table 1.** Benchmark scheduling rules for delay-sensitive traffic

| Strategy | Priority function |
|---|---|
| M-LWDF [12] | $\gamma_i \frac{\chi_{i,\varphi}^{(n)}}{R_{i,ave}^{(n)}} H_i^{(n)}$ |
| M-LWDFQ [39] | $\gamma_i \frac{\chi_{i,\varphi}^{(n)}}{R_{i,ave}^{(n)}} N_{Q_i}^{(n)}$ |
| EXP-PF [12] | $\gamma_i \frac{\chi_{i,\varphi}^{(n)}}{R_{i,ave}^{(n)}} \exp\left(\frac{\gamma_i H_i^{(n)} - \overline{\gamma_i H_i'}}{1+\sqrt{\overline{\gamma_i H_i'}}}\right)$ |
| EXP-PFQ [39] | $\gamma_i \frac{\chi_{i,\varphi}^{(n)}}{R_{i,ave}^{(n)}} \exp\left(\frac{\gamma_i N_{Q_i}^{(n)} - \overline{\gamma_i N_{Q_i}^{(n)}}}{1+\sqrt{\overline{\gamma_i N_{Q_i}^{(n)}}}}\right)$ |
| LOG-RULE [15] | $b_i \log\left[1.1 + \left(a_i \frac{H_i^{(n)}}{H_{i,max}^{(n)}}\right)\right] \chi_{i,\varphi}^{(n)}$ |
| EXP-RULE [15] | $b_i * exp\left[\frac{\left(a_i \frac{H_i^{(n)}}{H_{i,max}^{(n)}}\right)}{1+\sqrt{\overline{H_i'}}}\right] \chi_{i,\varphi}^{(n)}$ |

According to the table, $\gamma_i$ is a constant whose value is adjusted to account for different delay requirements for different flows. $N_{Q_i}^{(n)}$ is the number of packets residing in the queue of user i's flow at the eNodeB. In order to provide fairness, the delay-based rules (M-LWDF, EXP-PF, LOG-RULE, and EXP-RULE) use the HoL delay, whereas the queue-based rules (M-LWDFQ, EXP-PFQ) utilize the queue size of each flow. In the log (LOG-RULE) and exponential (EXP-RULE) rules, $b_i = \frac{1}{R_{i,ave}^{(n)}}$ and $a_i$ is a tunable parameter. The higher the value of $a_i$, the higher will be the priority of the delay-sensitive flows. All the priority rules shown in Table 1 are for delay-sensitive traffic. These rules calculate the priority for the best-effort traffic according to the PF rule given as:

$$\mathrm{PRF}_{i,\varphi}^{(n)} = \frac{\chi_{i,\varphi}^{(n)}}{R_{i,\mathrm{ave}}^{(n)}} \tag{21}$$

where time-averaged throughput $R_{i,\text{ave}}^{(n)}$ is defined as:

$$R_{i,\text{ave}}^{(n)} = R_{i,\text{ave}}^{(n-1)} * (1 - \frac{1}{n_w}) + \frac{1}{n_w} * R_i^{(n-1)}. \quad (22)$$

where $R_{i,\text{ave}}^{(n-1)}$ is the average throughput at scheduling instant *n*-1. $R_i^{(n-1)}$ is the number of bits transmitted at scheduling instant *n*-1. $n_w$ is the size of the time-average window also known as the exponential averaging constant. The higher the size of the time-average window, the higher the impact of the instantaneous channel quality.

## 4.2 Simulation scenario

In order to investigate the performance of the proposed scheduling algorithm, a link-level simulator [30, 31] built on MATLAB's object-oriented features is selected as the simulation platform with all the basic features of an LTE link layer. The fuzzy controllers are designed by utilizing the Matlab's Fuzzy Logic Toolbox. The fuzzy logic priority scheduling, under varying load, has been studied in [21], where only video traffic was considered. In this work, delay-sensitive traffic is characterized by video and VoIP flows. In order to simulate best-effort traffic, CBR flows with the data rate of 400 Kbps are selected. On the other hand, 64 Kbps traffic with the threshold packet loss rate of 1% and maximum delay budget of 100 ms is selected for VoIP users. These QoS parameters are selected according to LTE QCI (QoS class indicators) [32]. Video traffic is generated from a trace file [33] with the average and peak traffic rates of 530 and 1,500 kbps, respectively. The maximum delay budget for video packets is 200 ms whereas the threshold packet loss rate is 5%. For non-critical video applications, 5% packet loss rate corresponds to a peak signal-to-noise ratio (PSNR) of approximately 29 to 30 dB [34]. Therefore, 5% is considered as the threshold packet loss rate for an acceptable video quality. Table 2 reports the simulation parameters adopted for the LTE system and the wireless channel. We simulate 18 video flows (9.54 Mbps), 27 VoIP flows (1.728 Mbps), and 9 best-effort flows (3.6 Mbps) corresponding to a total average input traffic rate of 14.868 Mbps. The main motivations for such traffic distribution are the following:

- It has been reported in [35] that by 2015, approximately 66% of mobile's traffic (in terms of petabytes per month) will be video and the proportion of VoIP traffic will be a minority. Therefore, the proportion of traffic in our simulation scenario is dominated by video followed by the best-effort and VoIP traffic. Specifically, we selected a loaded network with 64% video, 11% VoIP, and 25% best-effort traffic (in terms of average input traffic at the eNodeB).

- The proposed scenario corresponds to an average input traffic rate of 14.868 Mbps. In order to evaluate the channel utilization in terms of average spectral efficiency, we simulate an optimum sum rate maximization strategy. The optimum strategy maximizes the system throughput without considering the delay constraints. The average channel quality (in terms of SINR) of the users is set such that the total system throughput, sum throughput of all the flows, produced by the throughput maximization strategy [36] is 13.6 Mbps (2.72 bits/s/Hz). This corresponds to a heavily loaded system where the input traffic is approximately 110%, in terms of bits/s/Hz, of the maximum system capacity. Our main goal is to study the fairness and efficiency performance of the proposed and benchmark scheduling rules when the delay bound and packet loss threshold constraints are considered.

We consider the time-averaged channel quality over the period, $n_c$ =100 ms. All the benchmark scheduling rules utilize the time-averaged throughput. In order to have a fair comparison, the exponential averaging constant $n_w$ is set to 100 ms. In the literature, the optimum size of the exponential averaging constant is from 100 to 1,000, with the 100 being utilized in scenarios yielding high fairness in terms of throughput.

**Table 2.** Simulation parameters - downlink LTE scheduling for delay-sensitive and best-effort traffic

| Parameters | Value |
|---|---|
| Simulator | LTE link level simulator [30,31] |
| Bandwidth, carrier frequency | 5 MHz, 2.1 GHz |
| UE distribution, cell radius | Uniform, 1 km |
| Channel | 3GPP-TU (typical urban) |
| Pathloss model | Hata-Cost-231 model |
| Shadowing model | Log-normal shadow fading |
| HARQ | Up to 3 synchronous retransmissions |
| Channel fading | Block fading (1 ms) |
| UE speed | 15 to 100 km/h (users moving independently at variable speed) |
| CQI averaging method | Mutual information effective SNR mapping |
| | (MIESM) [37] |
| $H_{max}$, $PLR_{thr}$ (video) | 200 ms, 5% [34,38] |
| $H_{max}$, $PLR_{thr}$ (VoIP) | 100 ms, 1% [32] |
| $H_{max}$, $R_{min}$ (best-effort) | 300 ms, 200 Kbps [32] |
| Number of video, VoIP and best-effort users | 18, 27, and 9 |
| Average rate requirements for video, | 530, 64, and 400 Kbps |
| VoIP and best-effort users | |
| $n_c$ (Time-averaged channel quality window) | 100 ms |
| and $n_w$ (Time-averaged throughput window) | |

The FCS scheduling strategy has the following tunable parameters:

- The time domain fairness parameter $\alpha_t$ mainly used to adjust the fairness level.
- The output fuzzy set for the DRC, proactive, and reactive controllers. As discussed in Section 3.3.1, the same output fuzzy set is utilized for all the controllers. The membership functions and fuzzy rules of the DRC are set such that by utilizing the same output fuzzy set for all the controllers, dynamic prioritization is achieved between the delay-sensitive and best-effort traffic.

Table 3 reports the settings considered in the simulations for the tunable parameters. The table reports three examples. In the first one (FCS1), the output fuzzy set is the same for both VoIP and video classes, i.e., video and VoIP traffic flows have the same prioritization. It is important to note that the LTE QoS architecture specifies QCI for each of the considered traffic classes [32]. According to the QCI, the VoIP traffic class has the highest priority followed by the video and best-effort traffic class. Therefore, in the second case (FCS2), VoIP is prioritized by increasing the maximum limit of the output fuzzy set from 2 to 2.2. The time domain parameter $\alpha_t$ is set to 2. Finally, in the third case, we retain the VoIP priority and increase the time-domain parameter from 2 to 3. The FCS2 and FCS3 modes follow the QCI service architecture where VoIP is prioritized over video traffic, while the FCS1 mode gives same priority to both the delay-sensitive traffic classes.

**Table 3.** Tunable parameters for FCS strategy

| Strategy | Output fuzzy set for video | Output fuzzy set for VoIP | $\alpha_t$ |
|---|---|---|---|
| FCS1 | {0, 2} | {0, 2} | 2 |
| FCS2 | {0, 2} | {0, 2.2} | 2 |
| FCS3 | {0, 2} | {0, 2.2} | 3 |

## 4.3 Results and discussion

First, we analyze the fairness and efficiency performance of the FCS strategy according to the settings reported in Table 3. Next, we compare the FCS strategy with the benchmark scheduling strategies reported in Section 4.1.Results in terms of packet loss rate (delay-sensitive flows) and throughput (best-effort flows) for the proposed rule with different parameters are shown in Figures 8 and 9, where users are arranged in a decreasing order of the channel quality. The user with lowest index has the best channel quality which then decreases with the increase in user index.Using the same prioritization (output fuzzy set) for video and VoIP (FCS1, Figure 8) results in a higher QoS violations for the VoIP flows, i.e., seven VoIP flows are violating the 1% PLR threshold. On the other hand, three video flows are violating the 5% PLR threshold.In the second set of simulations (FCS2, Figure 8), there is a significant reduction in the PLR of the VoIP flows, i.e., only one VoIP flow has a delay bound violation (PLR) of more than 1%. In FCS2 mode, the impact of the time-domain priority is higher for the VoIP flows. The increase in the HoL delay and PLR prioritizes VoIP flows more than the video flows. The result is an increase in the PLR of the video flows as shown in Figure 8. There is also a slight reduction in the throughput of the best-effort flows. Higher limit of the fuzzy set, 2.2, for VoIP traffic serves well according to the QoS architecture of LTE as it is the highest traffic priority class. Any further increase in the maximum limit of the fuzzy set for VoIP traffic will penalize the video and best-effort flows by serving VoIP traffic.

Next, we analyze the impact of the time-domain parameter $\alpha_t$. An increase in the time-domain priority parameter (FCS3, Figure 9) allocates relatively more resources to the worst channel flows since time-domain priority is a fuzzy function of the HoL packet delay, PLR, and time-averaged channel quality. It is important to note that the proportion of video traffic (18 flows with average rate requirements of 540 kbps) is high with respect to the VoIP traffic (27 flows with rate requirements of 64 kbps). Therefore, an increase in $\alpha_t$ results in a significant improvement for video flows as shown in

Figure 9. In other words, more resources are allocated to the lower channel quality video flows and as a result, their PLR is reduced at the expense of a slight increase in the PLR of VoIP flows. There is also a marginal increase in the PLR of good channel video flows. According to the figure (FCS3, Figure 9), the worst served video flow has a PLR of approximately 5.4% and the worst served VoIP flow suffers from a PLR of 1.6%. Thus, the FCS3 mode results in an improved fairness performance for the delay-sensitive flows. Under high load, the time-domain priority of the delay-sensitive flows will always be higher than best-effort flows. Therefore, increase in $\alpha_t$ will further enhance the priority difference and results in a reduction in the throughput of best-effort flows.

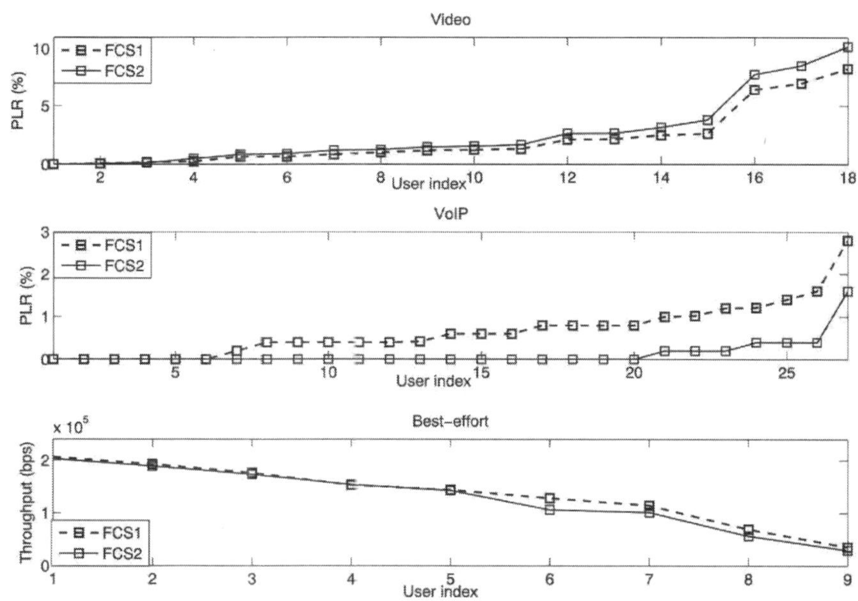


**Figure 8.** Performance of each traffic class with different output fuzzy set.

After analyzing the performance of the FCS rule for different design parameters, we compare the FCS rule with the well-known scheduling rules. First, we discuss the performance of state-of-the-art scheduling rules for delay-sensitive traffic. Figures 10 and 11 analyze the performance of state-of-the-art scheduling rules for delay-sensitive traffic and compare it with the FCS rule. Although M-LWDF is generally considered as the best scheduling rule for delay-sensitive traffic [12], the PLR performance of the M-LWDF scheduling rule for low channel quality video flows is very poor. According to Figure 10, the PLR of the worst served user is as high as 20% and approximately seven flows suffer from the QoS violation, i.e., having PLR above the 5% threshold. Higher QoS violations for video flows stem from the fact that the M-LWDF rule exploits time diversity by considering the time-averaged throughput. The video flows exhibit variable bit rate (higher peak-to-average rate ratio) characteristics. Therefore, the higher time-averaged throughput in the scheduling decision of the M-LWDF rule increases the probability of delay bound violations for the video flows having lower channel quality. Hence, high rate delay-sensitive flows with lower channel quality suffer from HoL delay bound violations. On the other hand, none of the VoIP flows suffer from delay bound violations (Figure 11) mainly because lower time-averaged throughput prioritizes the VoIP flows irrespective of their channel quality. M-LWDFQ reduces the QoS violations of the video flows by considering the queue size based on virtual token mechanism [39]. The PLR of the worst served user is approximately 12%, and there are only three flows violating the PLR threshold of 5%. The improved performance for video flows is mainly due the fact that the M-LWDFQ rule prioritizes high rate flows by considering the number of packets in the queue based on virtual token mechanism, as compared to the M-LWDF rule which relies on the HoL delay. As a result, flows having fewer packets in the queue are penalized if their channel quality is low. Figure 11 shows that seven VoIP flows have PLR of more than 1%. Therefore, the M-LWDFQ rule increases the QoS violation for the VoIP flows as compared to the M-LWDF scheduling rule.

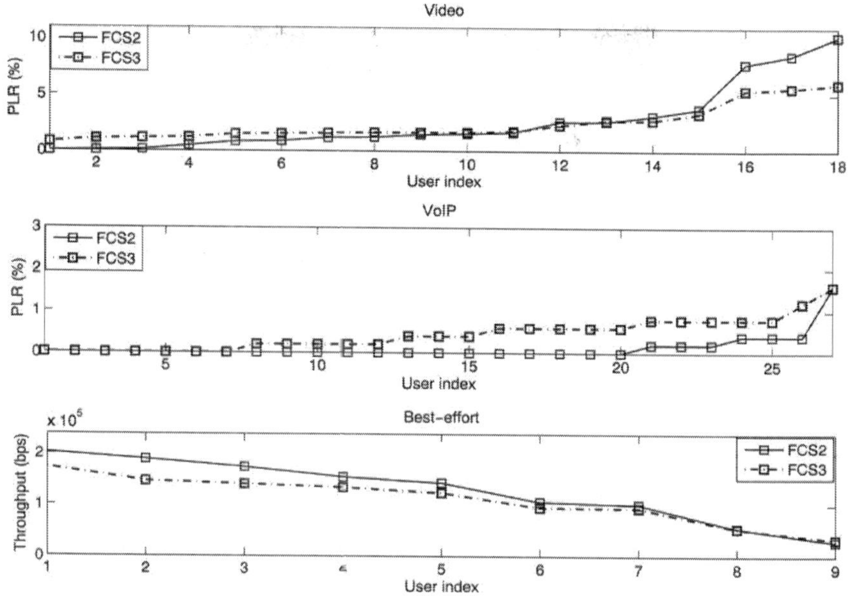


**Figure 9.** Performance of each traffic class with different time-domain priority.

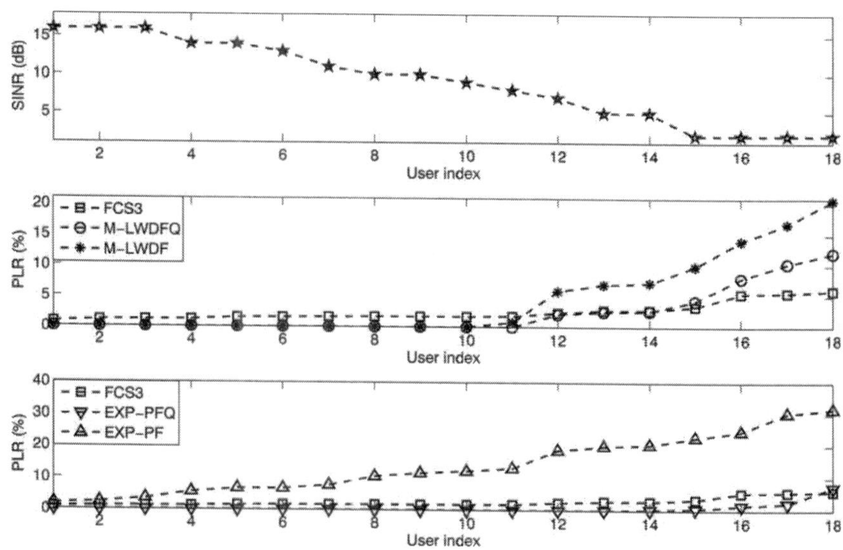


**Figure 10** Performance of the FCS, M-LWDF, M-LWDFQ, EXP-PF, and EXP-PFQ scheduling rules for the video flows.

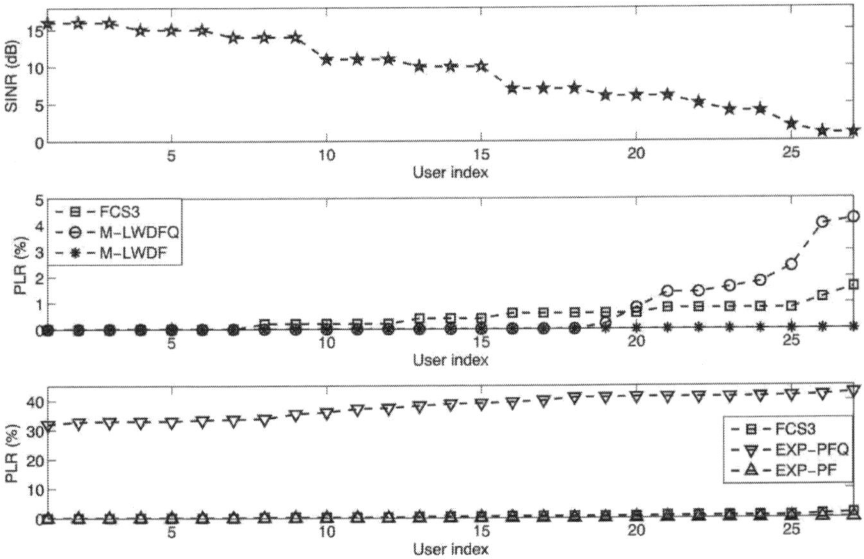


**Figure 11.** Performance of the FCS, M-LWDF, M-LWDFQ, EXP-PF, and EXP-PFQ scheduling rules for the VoIP flows.

When compared to the state-of-the-art scheduling rules, the FCS strategy improves the fairness performance for delay-sensitive flows mainly due to the fact that this scheduling rule considers the channel quality of a user in a novel way, by taking into account the past and current CQI feedbacks in the time domain priority metric. This allows the users with relatively low channel quality and high HoL delay to be prioritized in the time domain. As a result, the difference in the average waiting time of each flow's packet is low. On the other hand, state-of-the-art scheduling rules favor the good channel quality flows by serving them way before their packet's delay bound. These scheduling rules are highly unfair for the cell edge users as they require a substantial increase in the SINR of the cell edge users so that their packet's delay bound requirements are met. In the FCS scheduling strategy, the PLR over the moving average window is kept below the threshold for each of the delay-sensitive flows in the system. Therefore, this rule balance different flows' probabilities of QoS violations. It is important to note that the FCS strategy requires an admission controller to limit the arrival rate of delay-sensitive traffic within the achievable rate region. Since fairness is incorporated in the scheduling decisions, an increase in the arrival rate above the system capacity violates the QoS performance of the flows already being served.

The performance of the EXP-PF and EXP-PFQ scheduling rules for video and VoIP traffic classes is shown in Figures 10 and 11, respectively. For video flows, the EXP-PF scheduling rule increases the QoS violations significantly, i.e., approximately all the flows have delay violations of more than 5%. The performance of the EXP-PF rule for VoIP flows is the same as that of the M-LWDF rule. The M-LWDF and EXP-PF rule are delay-based schedulers. These scheduling rules prioritize VoIP flows mainly because of the lower rate requirements. The token-based version [39] of these scheduling rules penalizes the VoIP flows more because of the higher queue size of the video flows. EXP-PFQ performs worst for the VoIP flows mainly because the queue size of the VoIP flows always remains lower than the Video flows, which causes an exponential increase in the priority of the video flows. Therefore, all performance gain obtained by video flows penalizes to the VoIP and best-effort flows as shown in Figures 11 and 12. For best-effort flows (Figure 12), the performance of the EXP-PF rule is significantly better than other scheduling rules. The EXP-PFQ scheduling rule performs best for the video flows. However, it severely penalizes the VoIP flows.

All state-of-the-art scheduling rules prioritize best-effort flows by using the classical proportional fair Eq. 21. These rules prioritize delay-sensitive flows by using the linear, logarithmic, or exponential functions of the HoL delay as reported in Section 4.1. On the other hand, the FCS scheduling rule uses the same priority function for the best-effort and delay-sensitive flows, as given in Equation 19. The priority differentiation between the best-effort and delay-sensitive traffic classes is controlled by adapting the maximum limit of the output fuzzy set. The same priority function for each traffic class allows the exploitation of multi-user channel diversity across all the flows. This allows FCS rule to achieve intra-class and inter-class fairness which is not the case in state-of-the-art scheduling rules. The priority of the best-effort traffic class is dynamic and changes according to the QoS performance of the delay-sensitive flows.

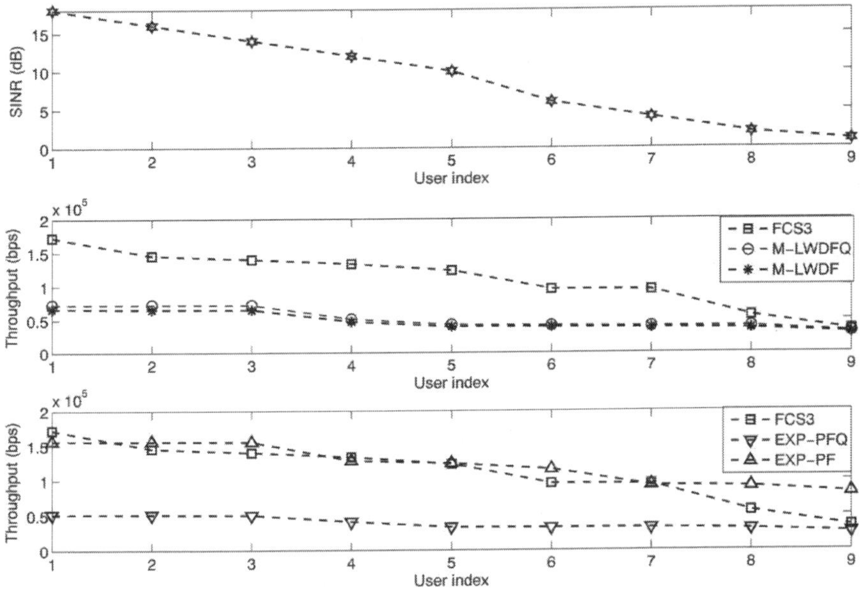


**Figure 12.** Performance of the FCS, M-LWDF, M-LWDFQ, EXP-PF, and EXP-PFQ scheduling rules for the best-effort flows.

The log-rule achieves the best performance for the best-effort flows but it is highly inefficient for the video flows as shown in Figure 13. It is important to note that the log rule uses the normalized HoL delay as the time-domain priority. The logarithmic variation of the normalized HoL delay has a marginal increase in the priority of the lower channel quality flows. For instance, if we set the tunable parameter $a_i$ to 50 (log-rule equation in Table 1) and analyze the priority difference between the 3 and 15 dB flows, the good channel quality flow with a normalized averaged delay of 0.3 results in a time domain priority of log[1.1+(50×0.3)]=2.78. On the other hand, the poor channel quality flow having normalized HoL delay of 0.9 results in a time-domain priority of log[1.1+(50×0.9)]=3.83. The logarithmic function marginalizes the priority of the poor channel quality flow as the delay urgency does not proportionately increases its priority. It is evident from Figure 13 that the log rule increases the PLR of the lower channel quality flows. The figure also shows the performance of the Exp-rule. According to the figure, the Exp-rule achieves better performance than the log-rule but it is highly unfair for the best-effort flows. The exponential function of the normalized HoL delay caters the delay urgency of the delay-sensitive flows better than the log rule. It is important to note that the FCS strategy increases the PLR of the VoIP flows as compared to the linear, logarithmic, and exponential delay based scheduling rules. However, the VoIP traffic class is packet loss tolerant and can tolerate the PLR threshold of 1%. The FCS strategy marginally violates the packet loss threshold of two VoIP flows, i.e., the worst served VoIP flows have a PLR of 1.2% and 1.6%.Figure 14summarizes the performance of all the scheduling rules. The figure reports the throughput achieved by each of the traffic classes. The last sub-figure shows the total system throughput, which is simply the sum of the throughput achieved by each of the traffic classes. When compared to the optimum channel utilization strategy, the FCS scheduler compromises approximately 10.5% of the total cell throughput while providing fairness and QoS provisions. It is clear from the figure that among all the aforementioned QoS-aware scheduling rules, the FCS scheduling rule achieves the best inter-class fairness in terms of the throughput achieved by each traffic class.

## 5 CONCLUSIONS

We proposed a composite scheduling strategy for downlink scheduling at the MAC layer for delay-sensitive traffic in wireless systems based on OFDMA. This strategy uses novel concept of providing fairness using fuzzy logic membership functions and its rule base, instead of relying on the rate based proportional fair strategies employed in the literature. Furthermore, we provide a framework for service class differentiation among different traffic classes by utilizing the fuzzy logic priority scheme. Our approach leads to a framework which provides intra-class as well as inter-class fairness. The design of the scheduling rule is robust, and it serves well in diverse channel and rate **requirements.**

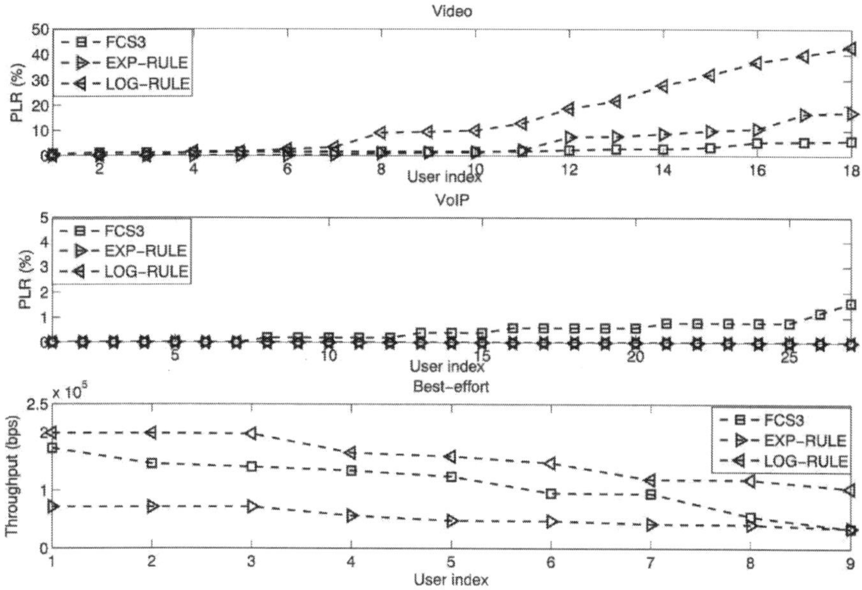


**Figure 13.** Performance comparison of the FCS strategy with the LOG and EXP scheduling rules.

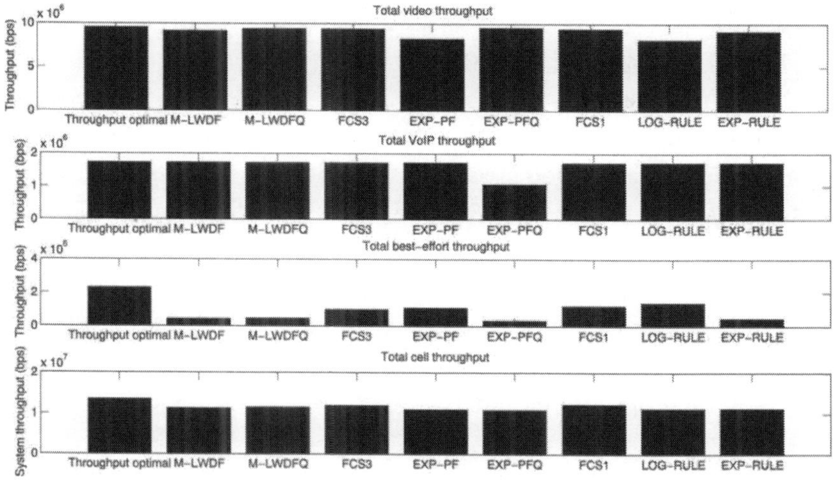


**Figure 14.** Throughput performance of all the considered scheduling rules.

## REFERENCES

1. Sesia S, Toufik I, Baker M: *LTE–The UMTS Long Term Evolution, From Theory to Practice*. Wiley, New york, USA; 2009.

2. Wong CY, Cheng RS, Letaief KB, Murch RD: Multicarrier OFDM with adaptive subcarrier, bit, and power allocation. *IEEE J. Selected Areas, Commun* 1999, 17(10):1747-1758.

3. Jang J, Lee KB: Transmit power adaptation for multiuser OFDM systems. *IEEE J. Selected Areas, Commun* 2003, 21(2):171-178.

4. Rhee W, Cioffi JM: Increasing in capacity of multiuser OFDM system using dynamic subchannel allocation. In *IEEE Vehicular Technology Conference (VTC)*. California, USA; 2000.

5. Jang J, Lee KB: Adaptive resource allocation in multiuser OFDM systems with proportional fairness. *IEEE Trans. Wireless Commun* 2003, 21(2):171-178.

6. Gross J, Karl H, Fitzek F, Wolisz A: *Intl. Conf. on Wireless Networks (ICWN)*. Las Vegas, USA; 2003.

7. Gross J, Klaue J, Karl H, Wolisz A: Subcarrier allocation for variable bit rate video streams in wireless OFDM systems. In*IEEE Vehicular Technology Conference (VTC)*. Florida, USA; 2003.

8. Khattab AKF, Elsayed KMF: Opportunistic subcarrier management for delay-sensitive traffic in OFDMA-based wireless multimedia networks. In *IST Mobile and Wireless Communications Summit*. Dresden, Germany; 2005.

9. Jalali A, Padovani R, Pankaj R: Data throughput of CDMA-HDR a high efficiency-high data rate personal communication wireless system. In *IEEE Vehicular Technology Conference (VTC)*. Tokyo, Japan; 2000.

10. Andrews M, Kumaran K, Ramanan K, Stolyar A, Whiting P, Vijayakumar R: Providing quality of service over a shared wireless link. *IEEE Commun. Mag* 2001, 39(2):150-154.

11. Rhee JH, Holtzman JM, Kim DK: Performance analysis of the adaptive EXP/PF channel scheduler in an AMC/TDM system. *IEEE Commun. Lett* 2004, 8(8):497-499.

12. Ramli HAM, Basukala R, Sandrasegaran K, Patachaianand R: Performance of well known packet scheduling algorithms in downlink 3GPP LTE system. In *IEEE Malaysia International Conference on Communications (MICC)*. Kuala Lumpur, Malaysia; 2009.

13. Capozzi F, Piro G, Grieco L, Boggia G, Camarda P: Downlink packet scheduling in LTE cellular networks: key design issues and a survey. *IEEE Commun. Surv. Tutorials* 2012, 15(2):1-23.

14. Khan N, Martini MG, Bharucha Z, Auer G: Opportunistic packet loss fair scheduling for delay-sensitive applications over LTE

systems. In *IEEE Wireless Communications and Networking Conference (WCNC)*. Paris, France; 2012.

15. Sadiq B, Madan R, Sampath A: Downlink scheduling for multiclass traffic in LTE. *EURASIP J. Wireless Commun. Netw* 2009., 2009:

16. Liu D, Lee YH: An efficient scheduling discipline for packet switching networks using earliest deadline first round robin. In *International Conference on Computer Communications and Networks (ICCCN)*. Dallas, USA; 2003.

17. Holtzman J: Asymptotic analysis of proportional fair algorithm. In *IEEE International Symposium on Personal, Indoor and Mobile Radio Communications (PIMRC)*. San Diego, USA; 2001.

18. Fodor G, Furuskar A, Skillermark P, Yang J: On the impact of uplink scheduling on intercell interference variation in MIMO OFDM systems. In *IEEE Wireless Communications and Networking Conference (WCNC)*. Budapest, Hungary; 2009.

19. Beh K, Armour S, Doufexi A: Joint time-frequency domain proportional fair scheduler with HARQ for 3GPP LTE systems. In *IEEE Vehicular Technology Conference (VTC)*. Calgary AB, Canada; 2008.

20. Jantzen J: Tutorial on fuzzy logic. Technical University of Denmark, Dept. of Automation, Tech. rep. 98-E-868 (1998)

21. Khan N, Martini MG, Staehle D: Opportunistic QoS-aware fair downlink scheduling for delay-sensitive applications using fuzzy reactive and proactive controllers. In *IEEE Vehicular Technology Conference (VTC)*. Las Vegas, USA; 2013.

22. Shin S, Ryu BH: Packet loss fair scheduling scheme for real-time traffic in OFDMA systems. *ETRI J* 2004, 26(5):391-396.

23. Koksal C, Kassab H, Balakrishnan H: An analysis of short term fairness in wireless media access protocol. In *ACM SIGMETRICS*. California, USA; 2000.

24. Garroppo RG, Giordano S, Iacono D, Tavanti L: Game theory and time utility functions for a radio aware scheduling algorithm for WiMAX networks. *Wireless Netw* 2011, 17(6):1441-1469.

25. Garroppo RG, Giordano S, Iacono D: Radio-aware scheduler for WiMAX systems based on time-utility function and game theory. In *IEEE Global Communications Conference (GLOBECOM)*. Hawaii, USA; 2009.

26. Ali S, Zeeshan M: A utility based resource allocation scheme with delay scheduler for LTE service-class support. In *IEEE Wireless Communications and Networking Conference (WCNC)*. Paris, France; 2012.

27. Ali S, Zeeshan M, Naveed A: A capacity and minimum guarantee-based service class-oriented scheduler for LTE networks. *Eur. J. Wireless Commun. Netw* 2013., 2013:

28. Monghal G, Pedersen KI, Kovács IZ, Mogensen PE: QoS oriented time and frequency domain packet schedulers for the UTRAN long term evolution. In *IEEE Vehicular Technology Conference (VTC)*. Singapore; 2008.

29. Assaad M, Mourad A: New frequency-time scheduling algorithms for 3GPP/LTE-like OFDMA air interface in the Downlink. In *IEEE Vehicular Technology Conference (VTC)*. Singapore; 2008.

30. Mehlfuhrer C, Wrulich M, Ikuno JC, Bosanska D, Rupp M: Simulating the long term evolution physical layer. In *European Signal Processing Conference (EUSIPCO)*. Glasgow, Scotland, UK; 2009.

31. Mehlfuhrer C, Ikuno JC, Simko M, Schwarz S, Wrulich M, Rupp M: The Vienna LTE simulators - enabling reproducibility in wireless

communications research. *EUR. J. Adv. Signal Process* 2011., 2011(29):

32. Tech. Specif. Group services and system aspects - policy and charging control architecture, 3GPP TS 23.203 V9.3.9, Release 9 (2009)

33. Seeling P, Reisslein M: Video transport evaluation with H.264 video traces. *IEEE Commun. Surv. Tutorials, in print*2012, 14(4):1142-1165.

34. Shin J, Kim JW, Kuo CCJ: Quality-of-service mapping mechanism for packet video in differentiated services network.*IEEE Trans. Multimedia* 2001, 3(2):219-231.

35. Cisco visual networking index: global mobile data traffic forecast update 2011-2016 *White Paper, Cisco* 2012.

36. Schwarz S, Mehlfuhrer C, Rupp M: Throughput maximizing multiuser scheduling with adjustable fairness. In *IEEE International Conference on Communications (ICC)*. Kyoto, Japan; 2011.

37. He X, Niu K, He Z, Lin J: Link layer abstraction in MIMO-OFDM system. In *International Workshop on Cross Layer Design (IWCLD)*. Jinan, China; 2007.

38. Khattab AKF, Elsayed KMF: Opportunistic scheduling of delay-sensitive traffic in OFDMA-based wireless networks. In*International Symposium on World of Wireless, Mobile and Multimedia Networks (WoWMoM)*. Buffalo-NY, USA; 2006.

39. Iturralde M, Wei A, Yahiya T, Beylot A: Performance study of multimedia services using virtual token mechanism for resource allocation in LTE networks. In *IEEE Vehicular Technology Conference (VTC)*. San Francisco, USA; 2011:1-5.

7

# TESTING AND ANALYSIS OF VOIPV6 (VOICE OVER INTERNET PROTOCOL V6) PERFORMANCE USING FREEBSD

**Asaad A. Abusin, M. D. Jahangir Alam, Junaidi Abdullah**

Faculty of Computing and Informatics (FCI), Multimedia University, Cyberjaya, Malaysia

## ABSTRACT

This study focuses on testing and quality measurement and analysis of VoIPv6 performance. A client, server codes were developed using FreeBSD. This is a step before analyzing the Architectures of VoIPv6 in the current internet in order for it to cope with IPv6 traffic transmission requirements in general and specifically voice traffic, which is being attracting the efforts of research, bodes currently. These tests were conducted in the application level without looking into the network level of the network. VoIPv6 performance tests were conducted in the current tunneled and native IPv6 aiming for better end-to-end VoIPv6 performance. The results obtained in this study were shown in deferent codec's for different bit rates in Kilo bits per second, which act as an indicator for the better performance of G.711 compared with the rest of the tested codes.

## KEYWORDS

VoIPv6 (Voice over Internet Protocol V6) Performance; Voice over Internet Protocol V6 Performance Testing; Voice over Internet Protocol V6 Performance Analysis; VoIPv6 Quality Testing in the Application Level


## 1. INTRODUCTION

In this study the G.711 was the codec technique in the VoIPv6 software that was used in both the VoIPv6 client and server couples in both the source IP and destination IP to investigate the voice packets traffic quality.

There are two main important and related considerations in selecting Audio Codec:

The delay that a codec will be introduced.

The Digital Signal Processor (DSP) speed that is required.

The DSP is measured in millions instructions per second (MIPS). Both factors affect the QoS of voice traffic, in addition to all that is the effect of the protocol performance [1,2].

## 2. LITERATURE REVIEW

With the great development of wireless communication technology and Internet, VoIP over wireless network is widely used. Providing QoS guarantees for VoIP applications is increasingly important, especially in mobile/ wireless networks due to their limited bandwidth and mobility [3].

The main difference between our study and [3] is that this study tests the VoIP performance in Mobile/Wireless Networks where as our study target the IPv6 protocol performance in wired networks, and as a holistic approach. In addition, [3] focused on studying VoIPv6 performance over Mobile networks whereas our study focus on the VoIPv6 performance in wired links.

The latest development from Cisco Systems is redefining the way businesses communicate [4]. Traffic analysis is essential to collect the statistical information about IP/UDP/RTP VoIPv6 packets such information like the arrival and inter arrival time of the voice packets, this will lead to better understanding the network and protocol performance, so that VoIPv6 networks with better QoS can be modeled. Our study encourages or pouch towards performing traffic analysis in networking systems to study the protocol performance.

IPv6 is documented in several RFCs (or request for comments) starting from RFC 2460. The IETF also publishes Experimental, Informational and Historic RFCs, and Best Current Practices. These RFC's specifies the V0IPv6 standards as mandates governing the protocol performance.

In [5], the main difference between this study and our study is that the testing tool in this study is Asterisk which is an open source/free software, whereas in our study the testing tool is FreeBSD which involve no overhead.

In [6] we are missing a capable test framework that is part of our main source tree.

With the stable release of FreeBSD 8.0 arriving last week we finally were able to put it up on the test bench and give it a thorough look over with the Phoronix Test Suite. We compared the FreeBSD 8.0 performance between it and the earlier FreeBSD 7.2 release along with Fedora 12 and Ubuntu 9.10 on the Linux side and then the Open Solaris 2010.02 b127 snapshot on the Sun OS side [6].

FreeBSD 8.0 introduced support for a TTY layer rewrite, network stack virtualization, improved support for the Sun ZFS file-system, the ULE kernel scheduler by default, a new USB stack, binary compatibility against Fedora 10, and improvements to its 64-bit kernel will allow a NVIDIA 64-bit FreeBSD driver by year's end, among a plethora of other changes. With today's benchmarking—compared to our initial Ubuntu 9.10 vs Free BSD 8.0 benchmarks from September—we are using the official build of FreeBSD 8.0 without any debugging options and we are also delivering a greater number of test results in this article, along with a greater number of operating systems being compared.

In recent years, Internet Protocol (IP) telephony has been a real alternative to the traditional Public Switched Telephone Networks (PSTN). IP telephony offers more flexibility in the implementation of new features and services. The Session Initiation Protocol (SIP) is becoming a popular signalling protocol for Voice over IP (VoIP) based applications. The SIP proxy server is a software application that provides call routing services by parsing and forwarding all the incoming SIP packets in an IP telephony network. The efficiency of this process can create large scale, highly reliable packet voice networks for service providers and enterprises. We established that the efficient design and implementation of the SIP proxy server architecture can enhance the performance characteristics of a SIP proxy server significantly. Since SIP proxy server performance can be characterised by its transaction states of each SIP session, we emulated the performance model of the SIP proxy server and studied some of the key performance benchmarks such as average response time to process the SIP calls, and mean number of SIP calls in the system. We showed its limitations, and then studied an alternative based SIP proxy server performance model with enhanced performance model and studied additional key performance characteristics such as server utilisation, queue size and memory utilization. In [7], they provided the comparative results between the predicted results with the experimental results conducted in a lab environment. The slit difference between this study in [7] and our study is that this study is a study of performance and scalability metrics of a SIP proxy server as a practical approach where as our study focus on the VoIPv6 performance in application level of IPv6 networks.

In [8], Mahani et al. study the effects of concurrent voice connections on the performance metrics of communication network such as queue length, waiting time, packets service time and is very important. Mathematical analysis of such network especially with long-tail traffic help for a good capacity planning and also lead to an accurate admission control algorithms.

In this study a mathematical model of a communication network supporting VoIP and back-ground traffic with long-tail service time is considered. Some problems of previous mathematical models are identified and a new queueing system is proposed in which specifically the coexisting of heavy-tail and voice flows is addressed. The long-tail service time is approximated via hyperErlang distribution and also to achieving an accurate performance model a Markov reward model is introduced. The available bandwidth for long-tail distribution varies according to the Markov chain, describing the utilisation factor of voice connection. Numerical results show a comparison between exponential and heavy-tail service time and finally the effects of concurrent voice connections on the service time of heavy-tailed background packets is shown.

This study [9] focus on the effects of concurrent voice connections on the performance metrics of communication network such as queue length, waiting time, packets service time and is very important. Mathematical analysis of such network especially with long-tail traffic will help us for a good capacity planning and also lead to an accurate admission control algorithms. In this study a mathematical model of a communication network supporting VoIP and back-ground traffic with long-tail service time is considered. Some problems of previous mathematical models are identified and a new queueing system is proposed in which specifically the coexisting of heavy-tail and voice flows is addressed. We are missing a capable test framework that is part of our main source tree.

This means that building-in testing when working on FreeBSD's base system requires extra steps, and so is harder than should be.

We currently keep our unit tests and regression test cases under/usr/src/tools/regression/. These tests use adhoc ways to build and execute their test cases. Test case reporting is alsonot standardized though some tests use the Perl Test Anything Protoco. Running these tests in an automated way (a test "tinderbox") is not always possible.

At this point of time we do not archive test logs. Even if we did so, analysis of historical test data would be tedious due to the ad-hoc nature of the test reports.

**The desirables from a FreeBSD test framework**

- The ability to write tests that cover all the functionality of the base system.
- The ability to manage multi-machine tests (i.e., distributed testing).
- A small, C-based test writing API that is easy to learn.
- Be capable of testing parts of the system that use threads.
- Integration with <bsd.*.mk>.
- Test logs should be easy to parse.
- Be available as open-source, and under a BSD-compatible license.

The rapid growth [1] of the Internet has led to the anticipated depletion of addresses in the current version of the Internet Protocol (IP), i.e., IPv4. This depletion has given rise to a newer version of the IP, i.e., IP version 6 (IPv6). IPv6 provides sufficient address space to meet the predicted increase of the Internet. Since IPv4 has already widely been deployed, it is required that the existing IPv4 and the newly added IPv6 can coexist and interoperate. Due to the incompatibility of the IPv4 and IPv6 headers, various mechanisms have been proposed to support the interoperability between IPv4 and IPv6. However, they are mostly designed for a static environment. Mobility support of mobile terminals in a mixed IPv4/IPv6 environment remains largely unexplored. It introduces additional overhead and delay to communications. In this paper, we analyze various handoff scenarios for a dual-stack mobile node with a predominant IPv6 home address roaming in a mixed IPv4/IPv6 environment. We investigate how handoffs can be supported and derive the handoff procedures for all scenarios. In addition, we analyze the impact of mobility support on the system performance in terms of handoff-signaling cost, handoff delay, and handoff-failure probability using our designed analytical models. Different traffic and mobility patterns are taken into account in the performance analysis. Numerical results are provided to demonstrate the performance of all handoff scenarios. Conclusions from this study can give great in-depth understanding and insights into designing new cost-effective mobility support mechanisms for IPv4/IPv6 transition and interoperability.

## 3. EXPERIMENTS

Depending on the above two considerations in the Introductions above, Tests were conducted on the audio codec's (G.711, G.721, G.723, and G.729) in two machines with different processor speed. An audio file of nine seconds duration time was used to test the encode and decode files of all the above mentioned audio codec standards. These tests were meant to know the execution time for the different codec's codes as a performance testing trial. These tests were conducted in machines with two different processor speeds, 200 MHz and 450 MHz, to examine the effect of the processor speed on the processing time of the codes.

The following PC specifications (comparatively old machines) were used to conduct the Codec tests:

Pentium 200 MHz MMX Processor 64 Mbytes 100 MHz Random Access Memory. FreeBSD 4.5 Release operating system. KAME version 2001OS28\FreeBSD [2,10,11].

# 4. RESULTS

Figures 1(a) and (b) show the time required to execute both G.711 encode and decode files in a PC with the above specifications. Figures 2(a) and (b) show the time required to execute both G.721 encode and decode files. Figures 3(a) and (b) show the time required to execute both G.723 encode and decode files, and Figures 4(a) and (b) show the time required to execute both G.729 encode and decode files.

From the above Figures 1-4 the best average execution time for both the encode and decode files is that of G.711, which act as an indicator for the better performance of G.711 compared with the rest of the tested codec's [2,10,11].

# 5. QUANTIFYING VOICE QUALITY

The voice quality was quantified and measured using a standardized ranking system called the Mean Opinion Score (MOS), which is a five-point scale described in ITU-T Recommendations P-800. On the surface, this system does not seem too scientific. The base of the MOS test is a matter of people listing to voice samples.

ITU-T P.800 makes number of recommendations regarding the selection of participants, the test environment, explanations to listeners, analysis of results. Different MOS tests performed on the same coding algorithm tend to give roughly similar results (**Table 1**).

Test results of the MOS values for different Codec. The MOS tests conclusion is coinciding with the result of Free BSD tests [11-13].

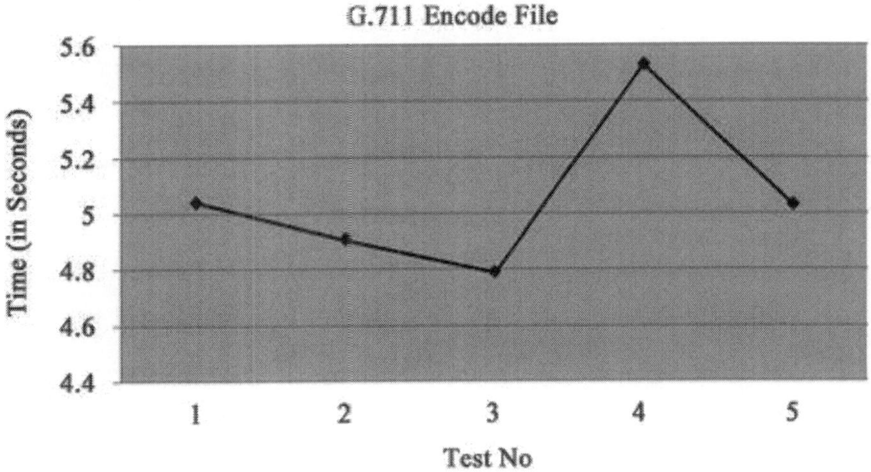


(a)

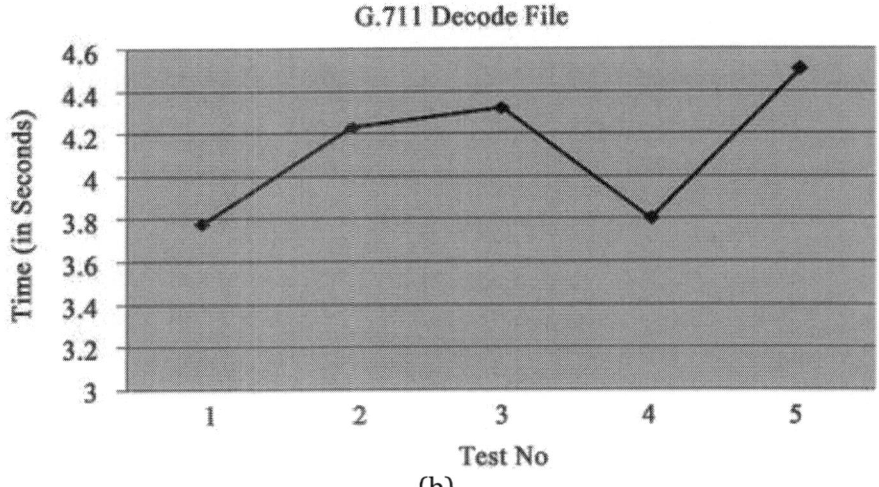


(b)

**Figure 1**. (a) G.711 encode file execution time average execution time is 5.065732; (b) G.711 decode file execution time average execution time is 4.126231 seconds.

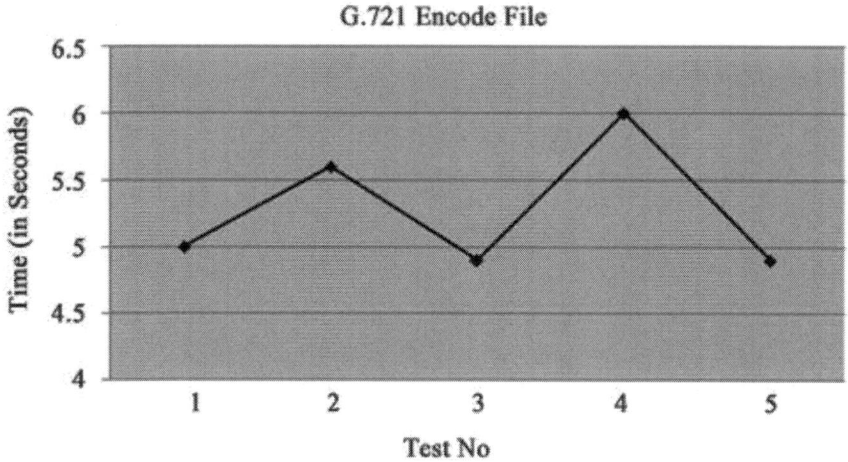


(a)

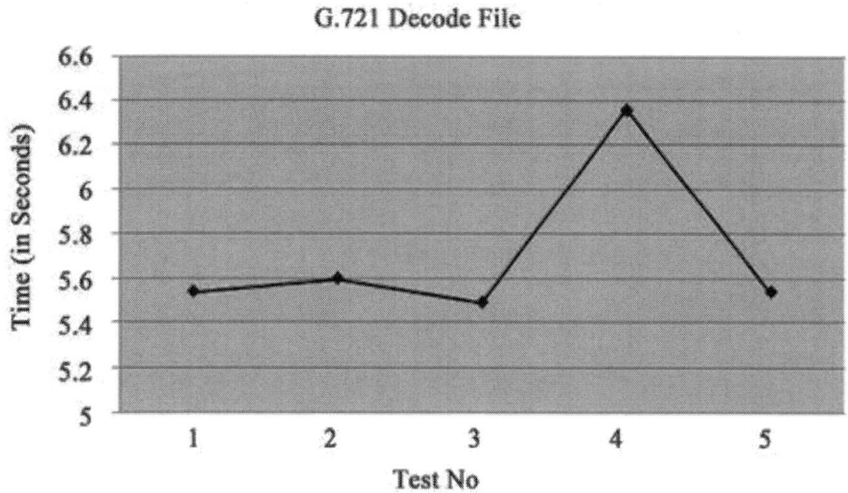


(b)

**Figure 2**. (a) G.721 encode file execution time average execution time is 5.267821 seconds; (b) G.721 decode file execution time average execution time is 5.709349 seconds.

IPv6 is documented in several RFCs (or request for comments) starting from RFC 2460. The IETF also publishes Experimental, Informational and Historic RFCs, and Best Current Practices (**Table 2**).

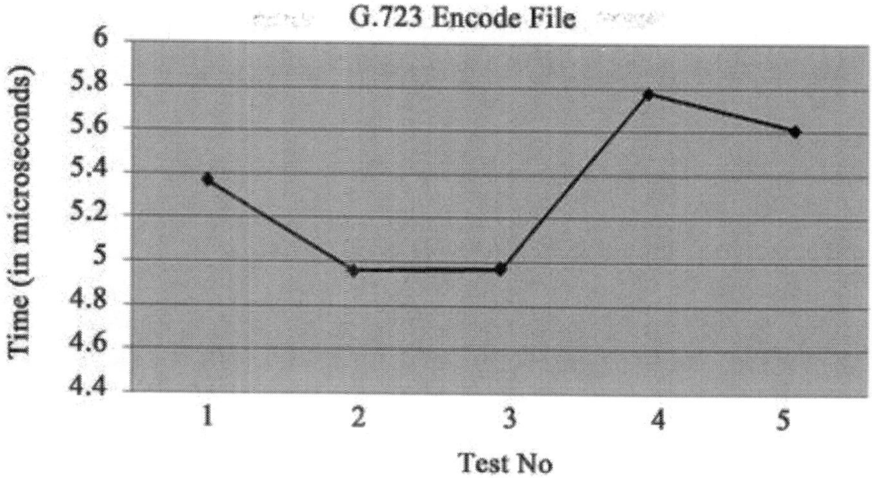

(a)

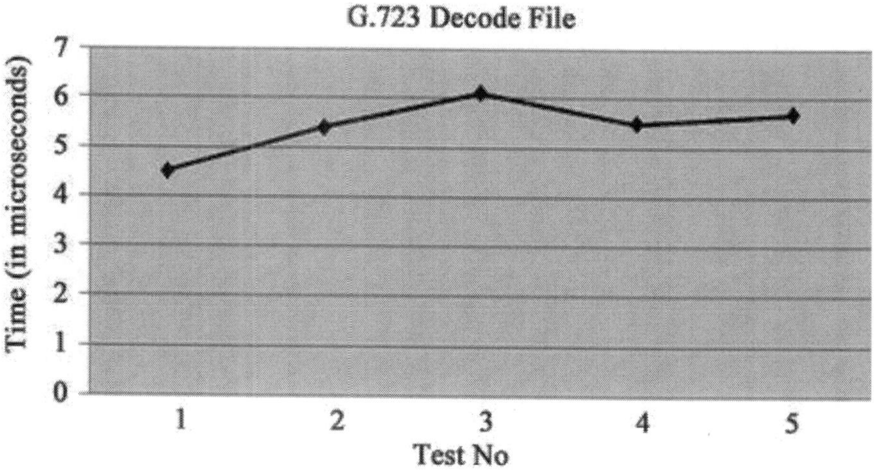

(b)

**Figure 3**. (a) G.723 encode file execution time average execution time is 5.338188 seconds; (b) G.723 decode file execution time average execution time is 5.421187 seconds.

The text of this document about the RFC Editor function is based upon the proposal that USC ISI submitted to the Internet Society in 2006. This proposal was to provide RFC Editor services during 2007-2008, with an optional extension to 2009 (this option was approved). Note that the proposal was written during the summer of 2006; many of the proposed tasks have in fact been completed [15].

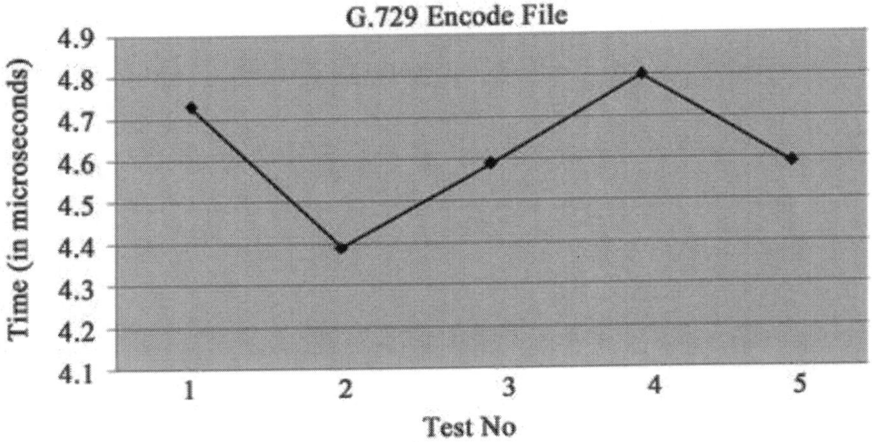


(a)

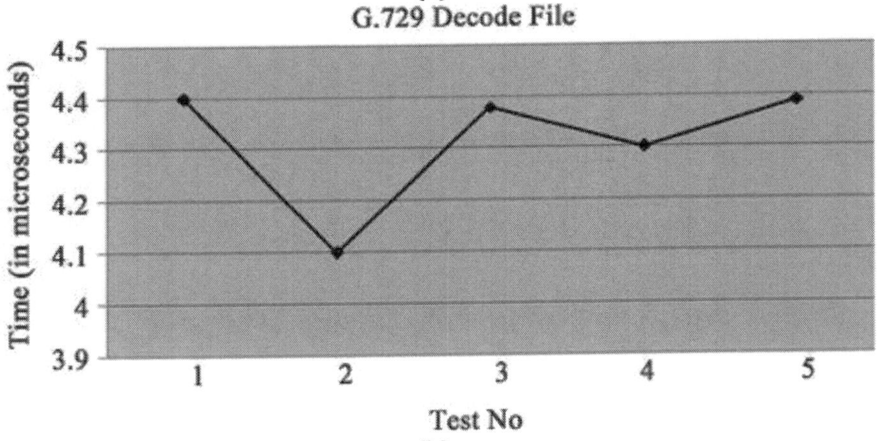


(b)

**Figure 4**. (a) G.729 encode file execution time average execution time is 4.617461 seconds; (b) G.729 decode file execution time average execution time is 4.309092 seconds.

**Table 1**. Test results of the MOS values for different Codec.

| Speech Coder | Bit Rate (Kbps) | MOS Value |
|---|---|---|
| G.711 | 64 | 4.3 |
| G.726 | 32 | 4.0 |
| G.723 | 63 | 3.8 |
| G.728 | 16 | 3.9 |
| G.729 | 8 | 4.0 |
| GSM Full Rate (RPE_LTP) | 13 | 3.7 |

## 6. CONCLUSIONS AND FUTURE WORKS

From the above Figures 1-4 we conclude the following:

- The best average execution time for both the Encode and decode files is that of G.711, which act as an indicator for the better performance of G.711 compared with the rest of the tested codes.

- Fluctuation and the degree of Uncertainty are high enough to affect the Quality of VoIPv6 performance.

- Any enhancement or quality improvement that could be made to the network infrastructure will lead directly to improvement in the VoIPv6 quality, Brood Band Internet is the best example of such improvement, this Brood Band Internet is been implemented in few countries in the world and Malaysia is expected to finalize this project by the beginning of the year 2012.

- Any enhancement or upgrading in the machines quality is expected to increase the overall quality of VoIPv6.

- • As future work these tests could be done in different platforms other than KAME for FreeBSD such as USAGI which is IPv6 implementation in Linux kernel, and Microsoft IPv6 stack implementation.

**Table 2**. These RFC's specifies the VoIPv6 standards as mandates governing the protocol performance [14].

| Protocol | Acronym | Purpose | RFC |
|---|---|---|---|
| Internet Protocol | IP | Physical network | RFC-791 |
| Internet Control Message Protocol | ICMP | Status messaging | RFC-792 |
| Transmission Control Protocol | TCP | Guaranteed delivery | RFC-793 |
| User Datagram Protocol | UDP | Coordination, Audio | RFC-768 |
| Telnet Protocol | TELNET | Remote login | RFC-764 |
| File Transfer Protocol | FTP | Network utility | RFC-765 |
| Simple Mail Transfer Protocol | SMTP | Email servers | RFC-788 |
| Network News Transfer Protocol | NNTP | Usenet | RFC-977 |
| Hypertext Transfer Protocol | HTTP | Web | RFC-2068 |

# REFERENCES


1. J. Xie, "Performance Analysis of Mobility Support in IPv4/IPv6 Mixed Wireless Networks," IEEE Transactions on Vehicular Technology, Vol. 59, No. 2, 2010, pp. 962-973.

2. V. Karthikeyan, N. Malmurugan and S. K. Jain, "Ambience-Based Voice over Internet Protocol Quality Testing Model," IETE Journal of Research, Vol. 55, No. 5, 2009, pp. 212-217.

3. H. Qian, "Measurement and Analysis of VoIPv6 QoS Parameters over Mobile/Wireless Networks," China Papers, 4 October 2010.

4. M. E. Flannagan, "Administering Cisco Qos in Ip Networks," Syngress, 11 April 2001, 519 p.

5. H. A. Ibrahim, "Open Source VoIPv6," Ministry of Communication & Information Technology (MCIT)-The Egyptian IPv6 Task Force, Cairo, 2008.

6. M. Larabel, "FreeBSD 8.0 Benchmarked against LINUX, Opensolaris," 2009. http://www.phoronix.com/scan.php?page=article&item=freebsd8_benchmarks&num=1

7. S. V. Subramanian and R. Dutta, "A Study of Performance and Scalability Metrics of a SIP Proxy Server—A Practical Approach," Journal of Computer & System Sciences, Vol. 77, No. 5, 2011, pp. 884-897.

8. A. Mahani, Y. S. Kavian, M. Naderi and H. F. Rashvand, "Heavy-Tail and Voice over Internet Protocol Traffic: Queuing Analysis for Performance Evaluation," IET Communications, Vol. 5, No. 18, 2011, pp. 2736-2743.

9. Integration TET-FreeBSD Testing Today, 2008. http:// wiki.freebsd.org/TetIntegration

10. S. W. M. Gimgibeom, "Performance Evaluation of VoIPv6 in a Large-Scaled u-Army Network," Korea Information and Communications Society, Chap. 36, No. 8, 2011, p. 897.

11. L. H. Duck and M. S. Won, "Network Performance of U-Army VoIPv6 Network Design Methodology Using Simulation," Korea Institute of Communication, Vol. 33, No. 10, 2008, p. 904.

12. A. N. Moldovan and C. H. Muntaean, "Towards Personalised and Adaptive Multimedia in M-Learning Systems," World Conference on E-Learning in Corporate, Government, Healthcare, and Higher Education, Hawaii, 18 October 2011, pp. 782-791.

13. P. Denisowski, "How Does It Sound," IEEE Spectrum, Vol. 38, No. 2, 2003, pp. 60-64.

14. "Internet Protocol Specification A," 1981. http://www.rfc-editor.org/rfc/rfc791.txt

15. S. Ginoza and A. Hagens, "The RFC Editor Function at ISI Bob Braden," USC Information Sciences Institute Marina del Rey, 2009. http://www.rfc-editor.org/RFCeditor.at.ISI.p

8

# EFFICIENT CONTROL CHANNEL RESOURCE ALLOCATION FOR VOIP IN OFDMA-BASED PACKET RADIO NETWORKS

**Yong Fan and Mikko Valkama**

Department of Communications Engineering, Tampere University of Technology, Tampere 33720, Finland

## ABSTRACT

We propose an efficient control channel resource allocation approach to enhance the performance of voice-over-IP (VoIP) in orthogonal frequency division multiple access- (OFDMA-) based next generation mobile communication systems. As the long-term evolution (LTE) of universal terrestrial radio access network (UTRAN), evolved UTRAN (E-UTRAN) is the first OFDMA-based packet radio network and thus selected in this paper as an application example. Our proposed physical downlink control channel (PDCCH) resource allocation approach for E-UTRAN is composed of bidirectional power control, inner loop link adaptation (ILLA), and outer loop link adaptation (OLLA) algorithms. Its effectiveness is validated through large-scale radio system level simulations, and simulation results confirm that VoIP capacity with dynamic scheduling can be further enhanced with this PDCCH resource allocation approach. Moreover, the VoIP performance requirements for international mobile telecommunications-advanced (IMT-Advanced) radio interface technologies can be met with dynamic scheduling together with proposed PDCCH resource allocation. Besides E-UTRAN, this approach can be introduced to other OFDMA-based mobile communication systems for VoIP performance enhancement as well.

# 1. INTRODUCTION

Mobile communication systems are gradually transforming from the systems mainly oriented to voice service to ones that can handle more high data rate services. Although data services are getting more momentum during this transformation, voice service still remains as the main source of revenue for mobile network operators [1]. For savings in capital expenditures and operating expenses, there is also a clear trend that all services are gradually being converged into packet switched (PS) domain. Related work [2, 3, 4] on supporting voice service through VoIP protocol in Third Generation cellular network, for example, high speed packet access (HSPA), have already been conducted. The future mobile networks will be completely IP-based and traditional circuit switched (CS) domain will not exist any more in them. Therefore, efficient VoIP support is a fundamental requirement for any emerging new systems.

The next generation mobile communication systems, for example, universal terrestrial radio access network (UTRAN) long-term evolution (LTE) and LTE-Advanced, worldwide interoperability for microwave access (WiMax), and ultra mobile broadband (UMB), all rely on orthogonal frequency division multiple access (OFDMA) as multiple access scheme. Conveying voice service through VoIP protocol in OFDMA-based mobile communication systems inevitably faces challenges caused by inherent VoIP traffic characteristics and stringent quality of service (QoS) criteria. VoIP traffic is low and constant bit-rate traffic, implying small size VoIP packets are transmitted at regular time intervals. The resource allocation for packet transmission is relatively small, thus necessitating multiuser sharing in frequency domain to effectively utilize available wide frequency bandwidth. Moreover, VoIP packets have to be scheduled frequently in order to satisfy the strict packet delay and loss based QoS criteria. In a typical packet radio system, such as evolved UTRAN (E-UTRAN), the transmission of each packet is done with dynamic scheduling by default [5] and evolved NodeB (eNB) needs to signal scheduled user equipment (UE) the resource allocation information through physical downlink control channel (PDCCH) every scheduling period, for example, one transmission time interval (TTI). As a result, the simultaneous scheduling of multiple users in a frequent manner demands lots of physical layer control signaling resources. The situation on control signaling consumption is further deteriorated by high VoIP capacity demand in PS radio network, which is supposed to accommodate a large amount of voice capacity currently handled in traditional CS mode.

Since the future mobile communication systems also target for high data rate transmission, it is preferable to allocate more radio resources for data transmission, thus leaving few resources for control signaling. Therefore, deploying VoIP in OFDMA-based packet radio systems encounters one unique challenge, that is, to satisfy high VoIP capacity demand with only limited physical layer control signaling resources. In E-UTRAN, PDCCH carries scheduling grants for both downlink and uplink, and its overall design was addressed in [6]. During the E-UTRAN standardization in Third Generation Partnership Project (3GPP), different solutions and schemes [7, 8, 9, 10, 11] have already been studied to overcome the limitations of PDCCH resources on VoIP capacity. All these aim at reducing the consumption of PDCCH resources and the possibility of conducting effective PDCCH resource allocation is not considered. The existing UTRAN LTE system level simulation results [9, 12] using simplified PDCCH modeling show that VoIP capacity with dynamic scheduling is limited by available PDCCH resources. The studies [13, 14] using realistic PDCCH modeling further confirm such limiting effect. Compared with delay-tolerant full buffer traffic, the impact of PDCCH limitation on delay-sensitive VoIP traffic is much more severe. While PDCCH limitation causes only marginal loss of cell throughput with full buffer traffic and there is also no evident relationship between cell throughput and the number of users scheduled per TTI [15], VoIP capacity in E-UTRAN is directly proportional to the number of users scheduled simultaneously in each TTI [13, 14]. On the other hand, it also implies great potential to significantly enhance VoIP capacity if effective PDCCH resource allocation is in place. However, the related study is obviously missing in available open literature.

To meet the ambitious performance target for international mobile telecommunication-advanced (IMT-Advanced) candidate radio interface technologies (RITs) [16], LTE is evolving into LTE-Advanced, and VoIP performance also needs to be improved. In the LTE-Advanced self-evaluation inside 3GPP, both dynamic scheduling and semipersistent scheduling are considered. While all existing solutions and schemes, such as dynamic packet bundling [9] and semipersistent scheduling [10, 11], try to reduce consumption of control channel resources, we consider a new paradigm in this paper to overcome limitation of control channel resources and further enhance VoIP performance, that is, to allocate limited control channel resources more effectively so that we can schedule more VoIP users dynamically in each TTI. Our proposed PDCCH resource allocation approach includes bidirectional power control, inner loop link adaptation (ILLA), and outer loop link adaptation (OLLA) algorithms. They are simple yet efficient and can be easily implemented requiring no changes to the existing 3GPP technical specifications.

The remainder of this paper is structured as follows. Section 2 gives a brief introduction of evolved universal terrestrial radio access (E-UTRA), the air interface of UTRAN LTE downlink as well as VoIP QoS and capacity criteria. Section 3 presents the design of proposed PDCCH resource allocation approach and related wideband channel quality indicator ( CQI ) measurement model and dynamic scheduling algorithm. Section 4 describes the overall simulation environment for VoIP on E-UTRA and used VoIP traffic model. The system simulation results in both 3GPP LTE and ITU IMT-Advanced scenarios are examined and analyzed in Section 5 before we draw final conclusions in Section 6.

# 2. VOIP ON E-UTRA

## 2.1. E-UTRA Overview

Figure 1 illustrates time-frequency resource structure of E-UTRA [17]. In time domain, the duration of one subframe, also called transmission time interval (TTI) is 1 ms, consisting of 2 equally sized time slots of 0.5 ms. Each time slot accommodates 6 or 7 OFDM symbols, depending on whether a short or extended cyclic prefix is applied. In each TTI, the first 1 to 3 OFDM symbols can be allocated for control signaling and the remaining OFDM symbols are used for data transmission. In this study, we assume that each TTI contains 14 OFDM symbols and the first 3 OFDM symbols are all reserved for control channels. In frequency domain, the minimum resource unit allocated for user data transmission is physical resource block (PRB), each containing 12 adjacent subcarriers with sub-carrier spacing of 15 KHz. In this study, we use a configuration of 5 MHz spectrum allocation with 25 PRBs. In time-frequency resource grid, resource element (RE) is the smallest time and frequency resource, spanning over 1 OFDM symbol in time domain and 1 sub-carrier in frequency domain. The minimum resource unit allocated to user for control signaling is control channel element (CCE) containing 36 REs which are distributed over time duration of 3 OFDM symbols and the whole frequency bandwidth. In 5 MHz spectrum allocation, the total number of CCEs is 20 which are assumed to be equally divided between downlink and uplink.

## 2.2. VoIP QoS and Capacity Criteria

For voice service, the one way end-to-end delay should be below 250 ms to guarantee relatively good voice quality rating [18]. The packet delay includes delay in evolved packet core (EPC) and E-UTRAN. Since the focus of this study is on E-UTRAN part, we deduct packet processing delay and propagation delay of approximately 100 ms in EPC, the delay budget left for E-UTRAN is around 150 ms.

We further assume that both end-users are E-UTRAN users, the tolerable delay for medium access control (MAC) queuing and scheduling as well as physical (PHY) transmission should be strictly within 80 ms. To further improve voice quality and better account for variability in network, 3GPP uses 50 ms air interface delay bound (one-way delay from eNB to UE) in LTE system performance evaluation [19]. According to satisfied user criterion (SUC) also specified in [19], a user is regarded as satisfied if 98% of packets for the user are received successfully within 50 ms air interface delay bound when monitored over the duration of the whole voice call. The VoIP system capacity is further defined as the number of users per cell when more than 95% of the users in the cell are satisfied.

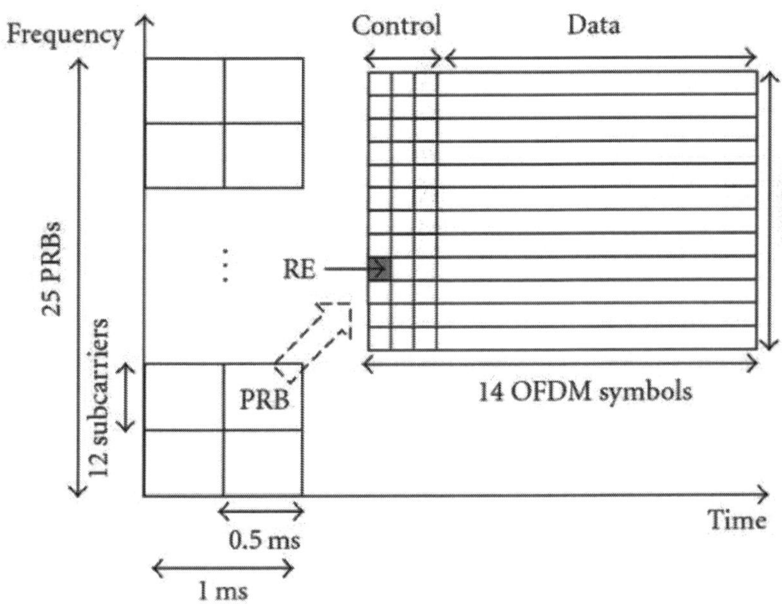


**Figure 1.** Time-frequency resource structure of E-UTRA.

# 3. PDCCH RESOURCE ALLOCATION AND RELATED RRM ALGORITHMS

Figure 2 illustrates the main radio resource management (RRM) entities located in eNB as well as interactions between them. Based on the buffer status, hybrid ARQ (HARQ) info, and used scheduling metric, dynamic packet scheduler generates a schedulable user list and then feeds it to PDCCH resource allocation unit for resource availability check. PDCCH resource allocation unit sequentially allocates control resources to user in the same order as that in the schedulable user list. Depending on the feedback from PDCCH resource allocation unit, dynamic scheduler selects users to be scheduled according to PDCCH resource constraints and allocate PRBs for user data transmission. Dynamic scheduler also consults physical downlink shared channel (PDSCH) inner loop link adaptation to determine the number of PRBs allocated to each user and MCS for each PRB. In this section, we will introduce the proposed PHY layer PDCCH resource allocation and closely related algorithms, such as CQI management and MAC layer dynamic packet scheduling.

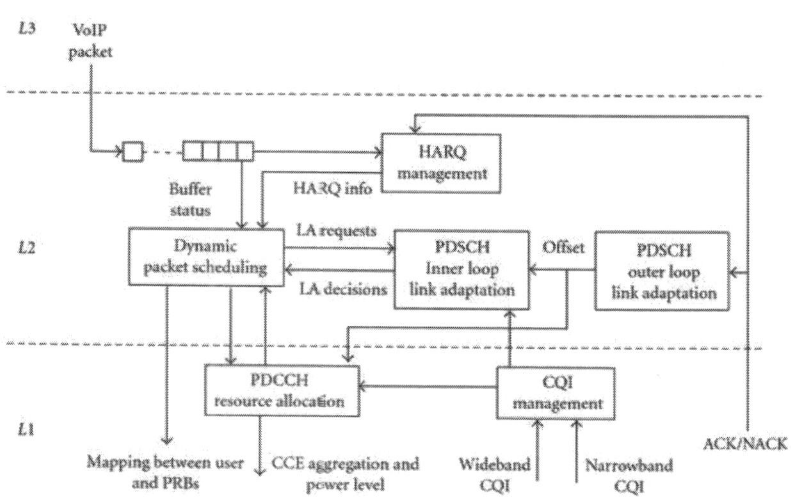


**Figure 2.** Main RRM functionalities and their interactions in eNB.

## 3.1. PDCCH Overview

PDCCH carries downlink scheduling assignments and uplink scheduling grants. The information fields in the scheduling assignments and scheduling grants include resource allocation information, transport format information, and HARQ information. Since there is no HARQ mechanism in PDCCH and the failure of PDCCH transmission will directly lead to failure of subsequent PDSCH transmission, the BLER target is set to 1% for the purpose of reliable PDCCH transmission. The MCS for transmission of PDCCH payload is QPSK modulation with four different effective coding rates of 2/3, 1/3, 1/6, and 1/12, corresponding to aggregation level of 1, 2, 4, 8 CCEs, respectively. A user in favorable radio channel conditions may require just 1 CCE with QPSK 2/3. Adaptive coding can be used to aggregate 2, 4, or even 8 CCEs with lower effective coding rate, thus improving the PDCCH coverage for users in worse radio channel conditions.

## 3.2. CQI Measurement Model

We determine the CCE aggregation level for each UE based on wideband CQI value as it reflects the average channel condition of UE. Wideband CQI is produced by taking the average of all ideal signal to interference noise ratio ( SINR ) values measured on each narrow subband across the whole frequency bandwidth. Therefore, the accuracy of wideband CQI modeling is greatly dependent on that of narrowband CQI modeling. The narrowbandCQI modeling follows the framework presented in [20]. Each UE periodically measures SINR on time duration $\Delta t$and subband bandwidth $\Delta f = N \cdot 180$ kHz, where $N$ is the consecutive number of PRBs within each measurement subband. To facilitate opportunistic scheduling in both time domain and frequency domain, $\Delta t$ and $\Delta f$ should be set to sufficiently small values so that frequency selectivity and time-variant behavior of radio channel can be well captured. Here, $\Delta t = 2$ ms and $\Delta f = 360$ kHz are used. The measured narrowband SINR is subject to an additive error due to limited number of pilot symbols within each subband. As the averaging value of narrowband SINR s, wideband SINR $SINR_{WB}$ is also subject to error, which is modeled as a zero-mean Gaussian random variable with standard deviation $\sigma_{WB} = 0.2$ dB. The measured realistic wideband SINR is then quantized to discrete CQI values ready for reporting. In UTRAN LTE downlink, the dynamic range of modulation and coding scheme (MCS) from quaternary phase shift keying (QPSK) with excessive coding to 64-quadrature amplitude modulation (64-QAM) with marginal coding is approximately 25 dB, which can be covered in 5 bits with quantization step size of $\Delta_{cqi} = 1$ dB. The generated wideband CQI $CQI_{WB}$ is expressed as

$$CQI_{WB} = \Delta_{cqi} \times \text{round}\left(\frac{SINR_{WB} + \sigma_{WB}}{\Delta_{cqi}}\right), \tag{1}$$

where round( ) is the operation of rounding to the nearest integer. Finally, the quantized wideband CQI s are reported to eNB with specified reporting schemes subject to certain feedback delay.

## 3.3. PDCCH Power Control and ILLA Algorithms

As seen in Figure 3, PDCCH resource allocation is composed of power control algorithm and link adaptation (LA) algorithm, which is further divided into ILLA and OLLA. PDCCH ILLA is responsible for estimating CCE aggregation level based on user's wideband CQI and adjustment offset from PDCCH OLLA algorithm. PDCCH power control is the key element in resource allocation. It determines the finally selected CCE aggregation and power level as well as balances allocation of power and CCE. Since power control and ILLA work closely with each other, we introduce them together here.

Previous study [15] on PDCCH resource allocation for full buffer traffic proposed a power boost algorithm. It artificially lowers CCE aggregation level by adding an initial power offset on user's wideband CQI before determining the required number of CCE. However, using high power offset will quickly run out of power while on the contrary CCE will easily be depleted if low power offset is applied. Therefore, it is hardly optimal in both power domain and CCE domain. In this study, we come up with a bidirectional power control algorithm to effectively balance the allocation of PDCCH power and CCEs, thus maximizing the number of dynamically scheduled users per TTI. The proposed PDCCH resource allocation can be divided into three steps and functions as follows.

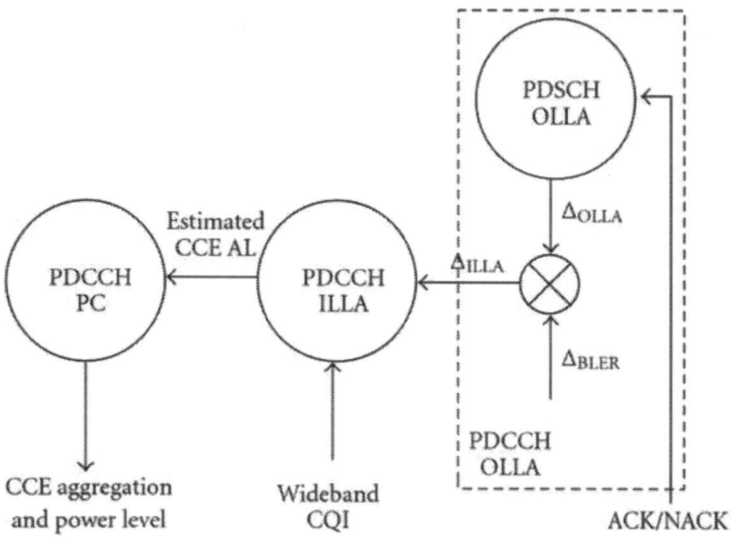


**Figure 3.** PDCCH resource allocation: power control and link adaptation.

***Step 1.*** Estimate CCE aggregation level: the estimated CCE AL $\tilde{N}_{CCE}$ is determined by choosing the minimum required number of CCEs satisfying equation $SINR_{WB} \geq SINR_{required,\ \tilde{N}_{CCE}}$, where $SINR_{WB}$ is user's wideband CQI taking into account measurement error and quantization effect as well as feedback delay, $SINR_{required,\ \tilde{N}_{CCE}}$ is the required SINR to reach the PDCCH target BLER of 1% with $\tilde{N}_{CCE}$.

***Step 2.*** Determine the actual CCE aggregation level and transmission power: without power control, the actual CCE AL would simply be $N_{CCE} = \tilde{N}_{CCE}$. The normal transmission power per sub-carrier would be $\overline{P}_{sc} = P_{eNB} / N_{sc}$, where $P_{eNB}$is the total eNB transmission power in mW, and $N_{sc}$ is the total number of subcarriers within the system bandwidth. Then, the PDCCH transmission power for user would be where $N_{RE}$ is the required number of REs for user, $N_{sym}$ is the number of OFDM symbols reserved for PDCCH per TTI, and $k$ is the size of one CCE in terms of the number of REs.

$$\left(P_{PDCCH} = \overline{P}_{sc} \cdot \frac{N_{RE}}{N_{sym}} = \overline{P}_{sc} \cdot \frac{N_{CCE} \cdot k}{N_{sym}}\right), \qquad (2)$$

With proposed power control algorithm, an individual power scaling factor $P_{SF}$ for user is applied on the top of normal sub-carrier transmission power $\overline{P}_{sc}$ to determine actual sub-carrier transmission power. The PDCCH transmission power for user

$$P_{PDCCH} = 10^{P_{SF}/10} \cdot \overline{P}_{sc} \cdot \frac{N_{CCE} \cdot k}{N_{sym}}. \qquad (3)$$

The algorithm to determine this user-specific power scaling factor $P_{SF}$ and actual CCE AL $N_{CCE}$ is illustrated using pseudocode below. For user consuming only one CCE, power reduction is conducted. For user consuming more than 1 CCE, either power reduction or boost is conducted depending on the additional amount of power needed to reach next lower CCE AL.

$P_{maxdecrease}$ is the maximum amount of power allowed to be reduced for each user and $P_{maxincrease}$ is the maximum amount of power allowed to be boosted for each user. Their values are set within the dynamics of eNB power amplifier and the absolute value of $P_{maxdecrease}$ is larger than that of $P_{maxincrease}$ to enable conservative power boost and aggressive power reduction.

**if** $\tilde{N}_{\text{CCE}} = 1$ **then**

$P_{\text{SF}} = \max\left(\text{SINR}_{\text{required},\,\tilde{N}_{\text{CCE}}} - \text{SINR}_{\text{WB}}, P_{\text{maxdecrease}}\right);$

$N_{\text{CCE}} = 1;$

**else if** $\tilde{N}_{\text{CCE}} = 2, 4, 8$ **then**

$P_{\text{SF}} = \text{SINR}_{\text{required},\,\tilde{N}_{\text{CCE}}/2} - \text{SINR}_{\text{WB}};$

**if** $P_{\text{SF}} < P_{\text{maxincrease}}$ **then**

$N_{\text{CCE}} = \tilde{N}_{\text{CCE}}/2;$

$P_{\text{SF}} = \text{SINR}_{\text{required},\,\tilde{N}_{\text{CCE}}/2} - \text{SINR}_{\text{WB}};$

**else if** $P_{\text{SF}} > P_{\text{maxincrease}}$ **then**

$N_{\text{CCE}} = \tilde{N}_{\text{CCE}};$

$P_{\text{SF}} = \text{SINR}_{\text{required},\,\tilde{N}_{\text{CCE}}} - \text{SINR}_{\text{WB}};$

**end if**

**end if**

(4)

***Step 3.*** Balance allocation of CCE and power: depending on whether there is still CCE or power left, the algorithm enters into the final step to balance CCE and power allocation. If there is power left, the algorithm will try to boost the power of user who originally cannot reach $\tilde{N}_{\text{CCE}}/2$ CCE in Step 2 due to the limitation from $P_{\text{maxincrease}}$. If there is CCE left, the algorithm will try to reduce the power of user who is originally boosted to $\tilde{N}_{\text{CCE}}/2$ CCE in Step 2.

## 3.4. PDCCH OLLA Algorithm

Since $CQI$ feedback from UE is subject to measurement error and reporting delay as well as other imperfections, OLLA algorithm is then employed to control the experienced average block error rate (BLER) for PDCCH transmissions. A classic OLLA algorithm follows the same principle of outer loop power control algorithm [21], and it relies on the success or failure of past transmissions to determine a compensation offset to ILLA algorithm. If an acknowledgement (ACK) is received for the first transmission, compensation offset factor $\Delta_{OLLA}$ is decreased by $\Delta down$ . On the contrary, if a negative acknowledgement (NACK) is received for the first transmission, compensation offset factor $\Delta_{OLLA}$ is increased by $\Delta up$ . The ratio of $\Delta up$ to $\Delta down$ determines the average BLER to which OLLA converges, as expressed in formula (5). There is one independent OLLA algorithm maintained for each user,

$$\mathrm{BLER} = \frac{1}{1 + (\Delta \mathrm{up}/\Delta \mathrm{down})}. \qquad (5)$$

Unlike PDSCH which can directly apply such type of OLLA, PDCCH cannot do so due to the lack of acknowledgment mechanism in PDCCH. Therefore, we rely on PDSCH OLLA to indirectly control the average BLER for PDCCH. The compensation offset factor $\Delta_{OLLA}$ generated from PDSCH OLLA is then scaled by a factor $\Delta_{BLER}$ to compensate the difference between BLER targets for PDSCH and PDCCH transmission. After that, the sum of these two factors is fed into PDCCH ILLA algorithm as $\Delta_{ILLA}$ for CCE aggregation level estimation.

## 3.5. Dynamic Scheduling Algorithm

A VoIP optimized dynamic scheduling algorithm [9] is used together with proposed PDCCH resource allocation method. The algorithm is separated into two parts: selecting users and assigning PRBs to users. In the user selection, we introduce a two-step scheduling candidate set (SCS) approach to efficiently control the interuser fairness as well as directly address the specific requirements for VoIP traffic scheduling.

For interuser fairness control, all the schedulable users on the basis of eNB buffer status are first sorted according to a relative wideband CQI metric and then organized into primary SCS. Instead of using traditional proportional fair metric based on user throughput, we come up with this new relative wideband CQI metric for interuser fairness control. The relative wideband CQI is defined as $CQI_{inst}(t) / CQI_{avg}(t)$, where $CQI_{inst}(t)$ is instantaneous wideband CQI and $CQI_{avg}(t)$ represents the average wideband CQI in the past. $CQI_{avg}(t)$ is calculated with traditional recursive method [22] by replacing user throughput with CQI value. Relative wideband CQI reflects user's instantaneous channel condition relative to the average and therefore indicates a favorable time instant for scheduling.

To address specific requirements for VoIP traffic scheduling, we further apply the following criteria of user classification on the primary SCS and users are then categorized into 3 groups in the secondary SCS for final user selection:

1) users with pending retransmission,

2) users whose delay of the oldest unsent packet in the eNB buffer is close to given threshold, and

3) remaining users other than the above two groups.

In PDSCH, there is no power control mechanism and each PRB is assigned with the same power level. The number of PRB allocated to each user is determined by adaptive approach based on user's channel condition. As for assigning PRBs to selected users, we give priority to first transmission users in choosing PRBs with relatively favorable channel quality and grant the retransmission users the priority to reserve the same amount of PRBs as in the original transmissions. By doing so, a high successful rate of first transmission can be achieved and then packet transmission delay over the air interface is reduced. Besides, retransmission users can anyway benefit from HARQ combining gain for correct reception, although they are given less freedom to choose suitable PRBs.

# 4. SYSTEM LEVEL SIMULATION SETUP

## 4.1. Overall Simulation Environment

We have developed a quasistatic system level simulator to investigate the effect of proposed PDCCH resource allocation on VoIP capacity enhancement. This simulation tool has been used extensively in generating various UTRAN LTE and LTE-Advanced performance evaluation results for 3GPP contributions and conforms to the guidelines specified in [19]. In the simulations, the related radio resource management algorithms shown in Figure 1 are all explicitly modeled. To reduce simulation time and facilitate easy comparison with existing results of VoIP on E-UTRA [9, 10, 12, 14], we consider 5 MHz system bandwidth configuration in all the deployment scenarios. The physical layer parameter settings are based on [17] and the physical layer performance of single radio link is obtained from separate link level simulations and fed to system level simulations through exponential effective SINR metric (EESM) [23]. The simulated network layout is a hexagonal grid composed of 19 cell sites with 3 sectors each. At the beginning of each simulation run, 2-Rx maximum ratio combining (MRC) receiver UEs are randomly placed in the simulation area and keep the position unchanged during this simulation run. Thus, distance-dependent path loss and shadow fading are constant while fast fading is updated every TTI according to the selected radio channel model, for example, typical urban (TU) 20 taps channel model for 3GPP LTE scenario. System level simulations are conducted with time resolution of one TTI, and several independent simulation runs are conducted to gather enough statistics. The main simulation assumptions are also summarized in Table 1.

**Table 1.** Main simulation parameters and values.

| Parameter | Value |
|---|---|
| Deployment scenario | 3GPP LTE: Macro Case 1 and 3<br>ITU IMT-A: indoor hotspot, urban microcell,<br>urban macrocell, and rural macrocell |
| HARQ | Asynchronous adaptive with chase combining<br>Num. of stop-and-wait processes: 8<br>Max. number of retransmissions: 3 |
| CQI | Measurement time duration: 2 ms<br>Measurement subband bandwidth: 360 KHz<br>Narrowband error standard deviation: 1 dB<br>Wideband error standard deviation: 0.2 dB<br>Quantization step: 1 dB<br>Feedback delay: 2 ms<br>Measurement period: 5 ms |
| PDSCH | Transmission power: 12 W (Macro scenario)<br>Num. of data OFDM symbols per TTI: 11<br>Num. of PRBs: 25<br>Num. of subcarriers: 300<br>BLER target: 10%<br>OLLA step up: 0.5 dB<br>OLLA step down: 0.05 dB<br>Max. OLLA step up: 3 dB<br>Max. OLLA step down: −1 dB |
| PDCCH | Transmission power: 8 W (Macro scenario)<br>Num. of control OFDM symbols per TTI: 3<br>Num. of CCEs (REs): 10 (360)<br>BLER target: 1%<br>Max. power increase: +2 dB<br>Max. power decrease: −4 dB |

## 4.2. VoIP Traffic Model

VoIP traffic is generated from adaptive multi rate (AMR) voice codec [24], which is widely used in 3GPP performance evaluation. Voice activity of user is modeled using 2-state Markov model recommended in [19]. In the case of using AMR 12.2 codec, voice encoder generates a voice frame of 31 bytes payload every 20 ms during activity periods. During silence periods, a silence descriptor (SID) frame of 5 bytes is sent for comfort noise generation purpose every 160 ms. Voice frame is encapsulated sequentially with network protocols including real-time transport protocol (RTP), user datagram protocol (UDP), and IP. After that, it is further encapsulated with radio protocols including packet data convergence protocol (PDCP), radio link control (RLC), and medium access control (MAC). The combined protocol header of 40 bytes for IPv4 will seriously degrade the air interface efficiency, thus necessitating robust header compression (RoHC). The total 9 bytes protocol overhead of compressed RTP/UDP/IP header [25] and PDCP/RLC/MAC header is added to each voice packet. The most important characteristics of VoIP traffic are summarized in Table 2.

**Table 2.** VoIP traffic characteristics.

| Parameter | Value |
|---|---|
| AMR voice codec rate | 12.2 kbps |
| Voice call length | 60 s |
| Voice activity factor | 50% |
| Mean of active period duration | 2 s |
| Mean of silence period duration | 2 s |
| Voice/SID packet size | 40 bytes/15 bytes |
| Voice/SID packet interarrival time | 20 ms/160 ms |

# 5. SIMULATION RESULTS ANALYSIS

This section presents simulated system performance of VoIP on E-UTRA in both 3GPP LTE and ITU IMT-A deployment scenarios. The VoIP capacity is evaluated according to SUC criteria described in Section 2.2. All the figures below are plotted with system load corresponding to 95% satisfaction point.

## 5.1. Capacity Enhancement with Proposed PDCCH Resource Allocation

The simulations are conducted for 3GPP Macro Case 1 and Case 3 deployment scenarios, of which the configurations are detailed in Table 3. Case 1 is an interference-limited scenario while Case 3 is an extremely noise-limited scenario in which the intersite distance is over 3 times that of Case 1. Besides enhanced performance with proposed PDCCH power control, baseline performance without PDCCH power control is also simulated for benchmarking purpose. Both without packet bundling and with packet bundling cases are considered. For packet bundling, a maximum of two packets is allowed to be sent together in one TTI to each UE.

**Table 3.** 3GPP LTE deployment scenarios.

| Deployment scenario | Macro case 1 | Macro Case 3 |
|---|---|---|
| Carrier frequency | 2 GHz | 2 GHz |
| Penetration loss | 20 dB | 20 dB |
| UE speed | 3 km/h | 3 km/h |
| Network layout | Hexagonal grid | Hexagonal grid |
| Intersite distance | 500 m | 1732 m |

### 5.1.1. VoIP Capacity Gain Analysis

Figures 4 and 5 summarize the achieved VoIP capacity and the relative capacity gain with PDCCH power control over without PDCCH power control. As shown in Figure 4, employing proposed PDCCH power control algorithm provides significant capacity gain over baseline performance without using PDCCH power control. The capacity gain ranges from 31% to 46%, depending on deployment scenario and whether packet bundling is enabled or not.

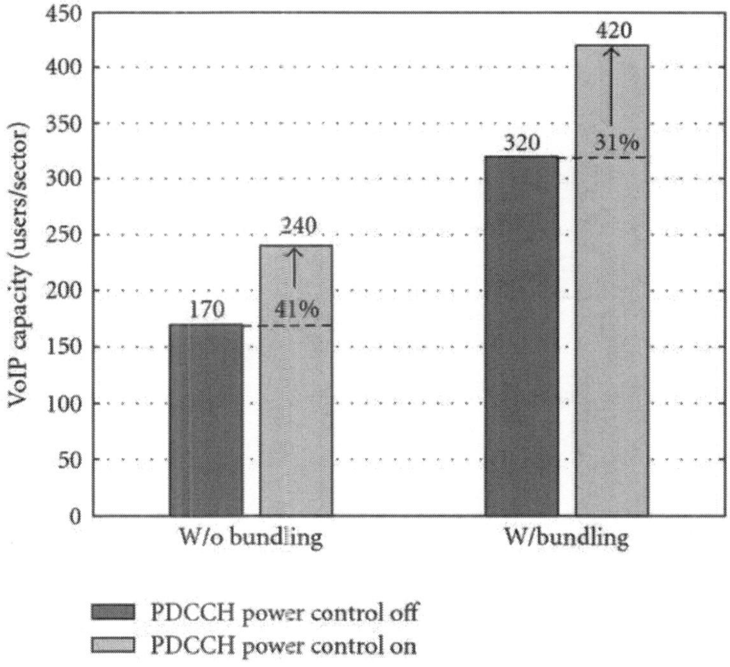


**Figure 4.** VoIP capacity comparison between without PDCCH power control and with PDCCH power control in 5 MHz system bandwidth (Macro Case 1).

One observation from Figures 4 and 5 is that power control gain in Case 3 is relatively higher than that in Case 1, either without bundling or with bundling. In Case 3, users experience lower SINR on average due to larger intersite distance and thus consume more PDCCH resources than those in Case 1. Therefore, users in Case 3 benefit more from PDCCH power control.

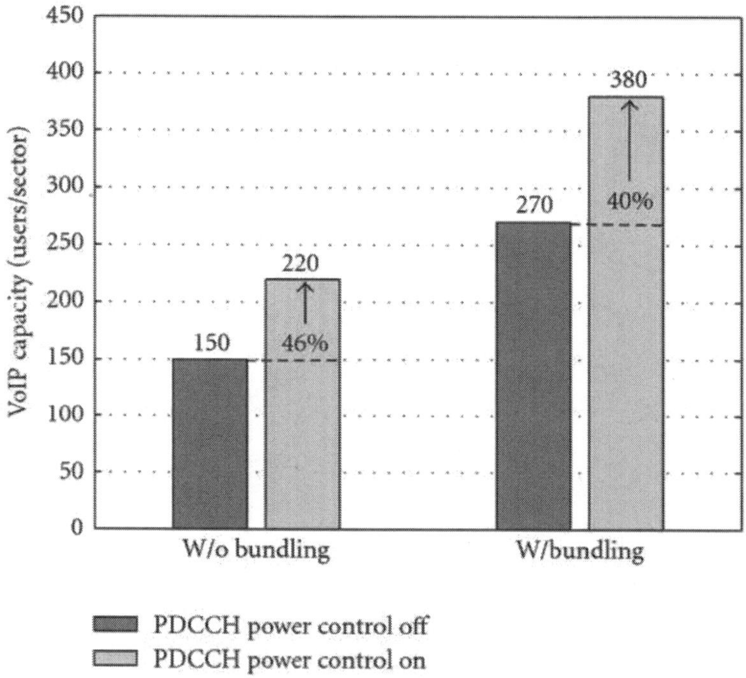


**Figure 5**. VoIP capacity comparison between without PDCCH power control and with PDCCH power control in 5 MHz system bandwidth (Macro Case 3).

The effectiveness of this power control algorithm is also validated from Figures 6 and 7. Figure 6 shows the distribution of CCE aggregation level per user, and Figure 7 shows the cumulative distribution function (CDF) of number of users scheduled per TTI, all using statistics collected from simulations in Case 3. As seen in Figure 6, CCE aggregation level of 4, 8 is reduced and aggregation level of 1, 2 is increased correspondingly with the aid of PDCCH power control. As a whole, the average aggregation level of user is lowered and the number of users scheduled per TTI is increased, which is also confirmed from Figure 7. As seen in Figure 7, the curve is shifted to the right side with the help from PDCCH power control, indicating the number of users scheduled per TTI is increased on average. In an OFDMA system, the increased number of VoIP users scheduled per TTI directly leads to increased VoIP capacity.

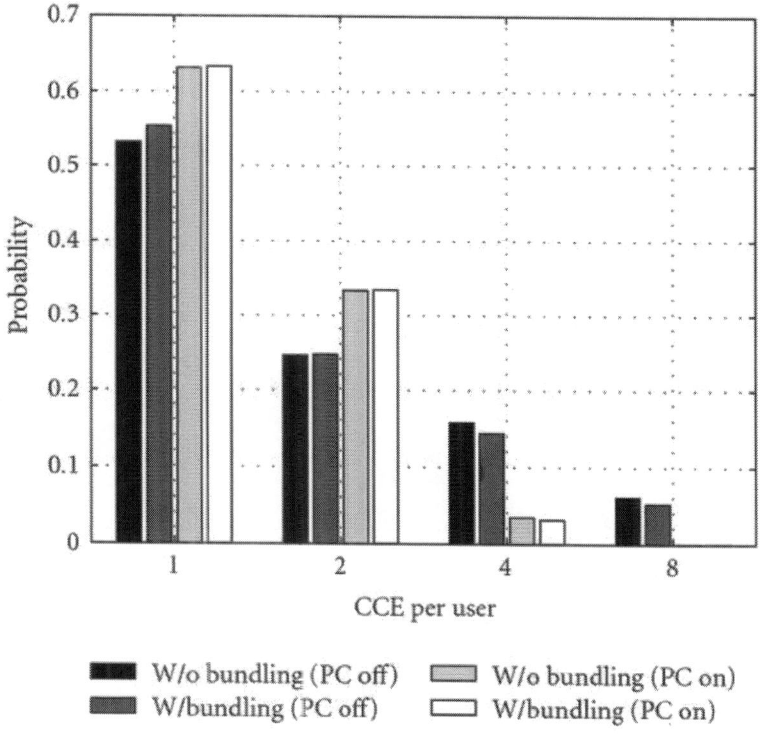


**Figure 6.** Distribution of CCE aggregation level (Macro Case 3).

Another observation from Figures 4 and 5 is the capacity gain from packet bundling is not compromised with proposed PDCCH power control. Our PDCCH resource allocation method strictly follows the calculated scheduling priority in user selection and does not prioritize the users in favorable channel conditions in PDCCH resource allocation. Therefore, users in favorable channel conditions always have enough time to accumulate two packets for bundling, and the packet bundling level can be guaranteed. It is also validated from Figure 8which shows the distribution of transmitted packets to each user per TTI. As seen in Figure 8, without power control, two-packet transmissions occupy around 60% of total transmissions in Case 3. With power control, the bundling level is even increased a bit.

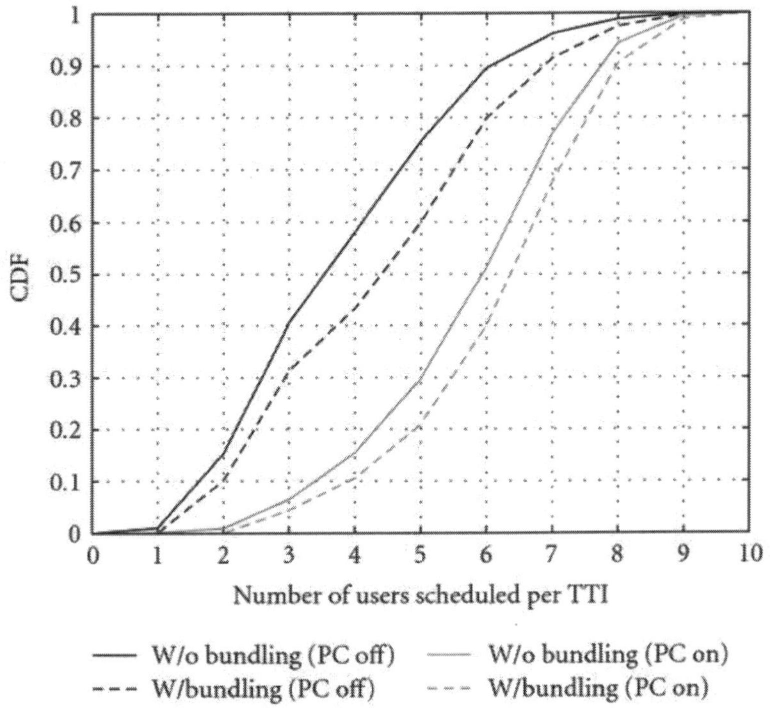


**Figure 7.** CDF of the number of users scheduled per TTI (Macro Case 3).

### 5.1.2. PDCCH and PDSCH Resource Utilization Analysis

Figure 9 shows the CDF of the amount of PDCCH resources utilized per TTI, in terms of RE and power. Without power control, the transmission power for each CCE is at the same level and PDCCH resource allocation is conducted actually only on one dimension:

CCE. Power is allocated in discrete quantities following the number of allocated CCEs. As seen in Figure 9, the shape of CDF of the number of REs used per TTI is exactly the same as that of CDF of the amount of PDCCH power used per TTI. Such kind of plain power allocation obviously leads to the waste of power resources in PDCCH. Therefore, PDCCH power is wasted mostly on users consuming 8, 4, or even 2 CCEs.

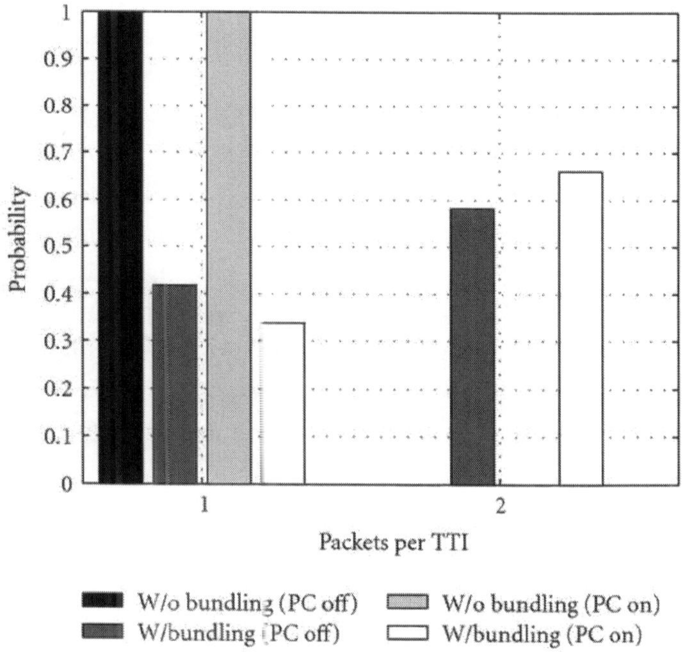


**Figure 8.** Distribution of the number of transmitted packets to each user per TTI (Macro Case 3).

With proposed PDCCH power control, the resource allocation is conducted on two dimensions: CCE and power. CCE is still allocated in discrete quantities, but power is instead allocated in continuous quantities, as can be observed from Figure 9. The over-allocated power on user with CCE aggregation level of 8, 4, 2 is used to exchange for reduction of CCE aggregation level to 4, 2, 1. The saved power from users consuming 1 CCE can also be used for this purpose. As a result, not only the total number of CCEs used per TTI can be decreased, but also the assigned power per TTI can be reduced. As also observed from Figure 9, the amount of used power and CCEs with power control is well below that without power control, although the achieved capacity is 40–46% higher. The average utilization rate of CCEs and power is approximately 80%, indicating the efficient balance between CCE and power allocation can also be achieved with this algorithm.

Figure 10 shows the CDF of the number of PRBs scheduled per TTI. Without power control, the average utilization rate of PRBs is only 40% even with the help from packet bundling. More than half of the PDSCH resource can not be efficiently exploited as there are not enough CCEs to schedule unused PRBs. With proposed power control, the average aggregation level of CCE is lowered and more users can be scheduled per TTI, leading to more efficient exploitation of PDSCH resources. As observed, the average utilization rate increases to 80% with power control. There is still space to increase VoIP capacity with dynamic scheduling by using transmit diversity in eNB and interference rejection combining (IRC) receiver in UE to further increase SINRand reduce the amount of PDCCH resources required by UE.

### 5.1.3. VoIP Packet Delay Analysis

Figure 11 shows the CDF of experienced packet delay for VoIP user. As observed, employing proposed power control does not deteriorate the delay performance. VoIP packet delay is mainly composed of two parts: queuing delay in eNB buffer and transmission delay over the air. With proposed power control, more users can be scheduled per TTI and the queuing delay in eNB buffer is reduced. Although the system load with power control is 46% higher than that without power control, the curve for without bundling is only slightly shifted to right side, implying minor increase in overall packet delay experienced by user. It is also the same case with bundling.

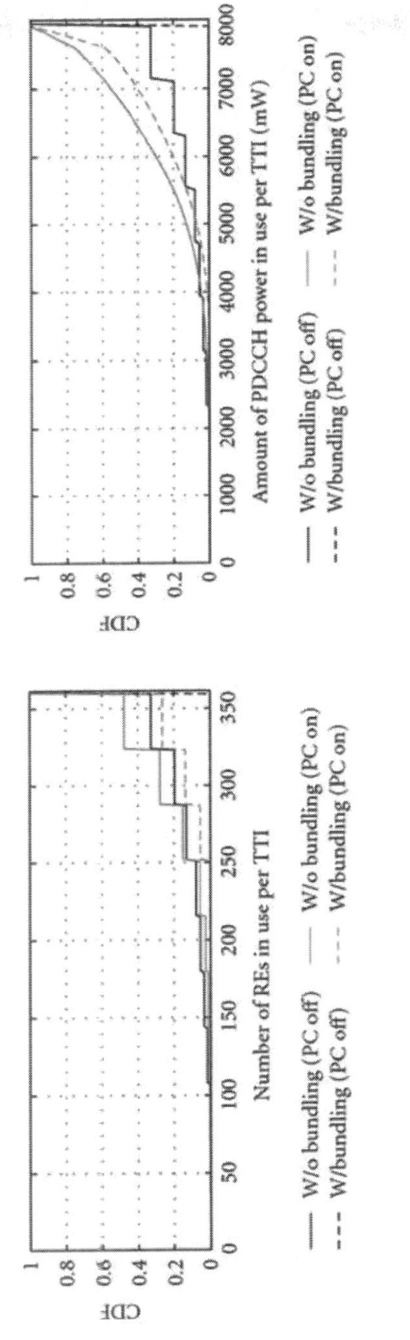


**Figure 9.** CDF of utilization rate of PDCCH CCE and power per TTI (Macro Case 3).

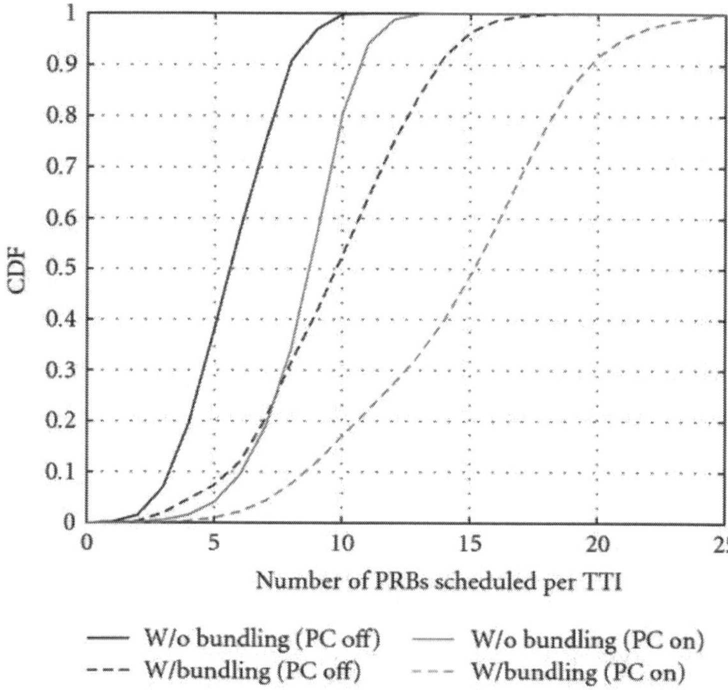


**Figure 10.** CDF of the number of PRBs scheduled per TTI (Macro Case 3).

## 5.2. Capacity Evaluation for IMT-Advanced

IMT-Advanced candidate RITs are required to be applied in a variety of deployment scenarios and support a range of environments. ITU deployment scenarios chosen by 3GPP for LTE-Advanced self-evaluation include indoor hotspot (InH), urban microcell (UMi), urban macrocell (UMa), and rural macrocell (RMa). The network layout, carrier frequency, intersite distance, and UE speed of interest in these scenarios are shortlisted in Table 4.

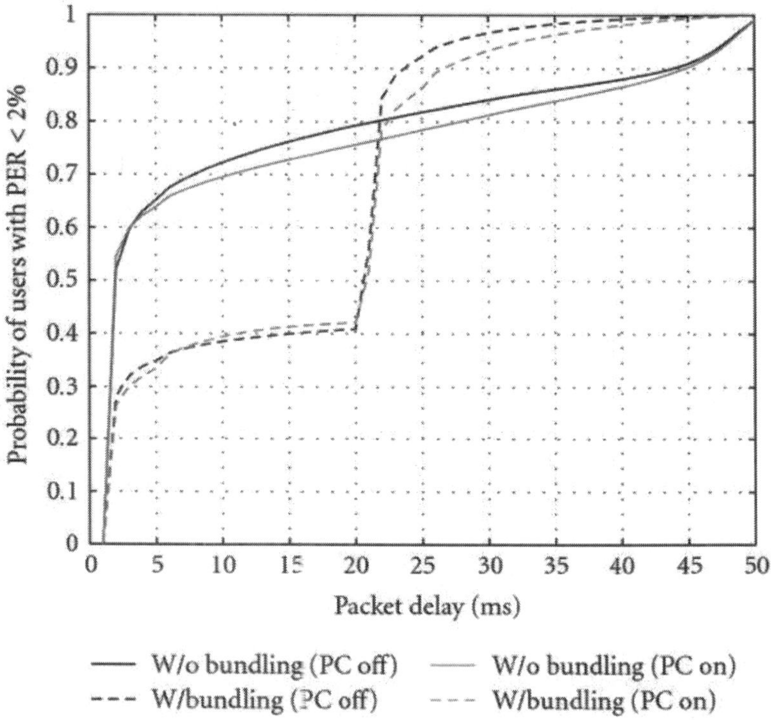


**Figure 11.** CDF of packet delay (Macro Case 3).

The simulations are conducted using the same quasistatic system level simulator, according to guidelines for evaluation of radio interface technologies for IMT-Advanced [26] as well as additional modeling and assumptions for different ITU deployment scenarios [27]. ITU simulation methodology is slightly different from 3GPP counterpart mainly in radio channel model and link budget setting. More details can be found in [26, 27]. Except that InH scenarios use a two-isolated cell layout, all the other three scenarios use the same simulated network layout as that used in 3GPP LTE simulations. VoIP traffic model, QoS criteria, link-to-system mapping model, and other simulation assumptions related to HARQ and CQI are the same as those used in the simulation of 3GPP LTE scenarios.

**Table 4.** ITU IMT-Advanced deployment scenarios.

| Test Environment | Indoor | Microcellular | Base coverage urban | High speed |
|---|---|---|---|---|
| Deployment scenario | Indoor hotspot | Urban microcell | Urban macrocell | Rural macrocell |
| Carrier frequency | 3.4 GHz | 2.5 GHz | 2 GHz | 800 MHz |
| Layout | Indoor floor | Hexagonal grid | Hexagonal grid | Hexagonal grid |
| Intersite distance | 60 m | 200 m | 500 m | 1732 m |
| UE speed of interest | 3 km/h | 3 km/h | 30 km/h | 120 km/h |

Table 5 summarizes the simulated VoIP capacity against IMT-Advanced requirements. Simulation results show enhanced VoIP capacity exceeds the requirements in all ITU test environments. It is concluded that IMT-Advanced requirements on VoIP capacity can be fulfilled with dynamic scheduling with proposed PDCCH resource allocation. Among these four scenarios, InH scenario includes only two-isolated cells with pedestrian users. UMi scenario has small intersite distance and includes pedestrian users only. Thus, high SINR and accurate CQI can be achieved in these two scenarios. PDCCH resource demand per user is not high and more users can be dynamically scheduled per TTI. In addition, frequency domain scheduling can be conducted efficiently and dynamic packet bundling can be utilized as much as possible. All these lead to high VoIP capacity. UMa scenario has slightly larger distance than that in InH and UMi scenarios and includes fast vehicular users. At speed of 30 km/h, the accuracy of CQI reported from UE begins to be vulnerable to fast changing radio channel conditions and gain from frequency domain scheduling and packet bundling might be reduced. On the other hand, due to relatively higher eNB transmission power in macro cell scenarios, the reserved higher PDCCH power leaves more room for efficient PDCCH power control and thus helps to boost VoIP capacity. RMa scenario has the largest intersite distance among all four scenarios and high-speed vehicular users. The frequency domain scheduling gain is lost due to inaccurate CQI in fast changing radio channel condition. But PDCCH power can take advantage of relatively higher PDCCH power in macro cell scenario. In addition, PDCCH OLLA is capable of partly compensating for the CQI errors. All these help to sustain the capacity in lower SINR and control resource demanding situation.

## 6. CONCLUSIONS

In this paper, we studied PDCCH resource allocation method oriented for VoIP optimized dynamic scheduling. A comprehensive set of system level simulation results were demonstrated, covering both 3GPP LTE scenario and ITU IMT-Advanced scenario. A thorough analysis of capacity gain, PDCCH and PDSCH utilization rate, and packet delay performance were provided. Postsimulation analysis further confirms the effectiveness of proposed PDCCH resource allocation method. The proposed bidirectional power control algorithm can efficiently balance the allocation of CCE and power in PDCCH and maximize the number of users scheduled per TTI, thus greatly improving the capacity of VoIP with dynamic scheduling. The relative capacity gain is approximately 31–46% over that without power control, depending on deployment scenario and whether packet bundling is enabled or not. Besides, PDCCH resource allocation does not impose negative impact on packet bundling level and packet delay performance. Moreover, in all four ITU IMT-Advanced deployment scenarios, enhanced VoIP capacity exceeds the target set in IMT-Advanced. This approach can also be applied in semipersistent scheduling as the retransmissions and SID transmissions in semipersistent scheduling also needs control signaling. Although we use UTRAN LTE downlink as an application example in this paper, the proposed PDCCH resource allocation method can be naturally applied to UTRAN LTE uplink as well as other OFDMA-based mobile communication systems for VoIP performance enhancement.

**Table 5.** VoIP capacity requirement in IMT-Advanced and simulated results.

| Test scenario | Minimum required capacity (Active users/sector/MHz) | Simulated capacity (Active users/sector/MHz) |
|---|---|---|
| InH | 50 | 105 |
| UMi | 40 | 94 |
| UMa | 40 | 80 |
| RMa | 30 | 72 |

## ACKNOWLEDGMENT

The authors appreciate the support from many colleagues, especially Markku Kuusela from Nokia Devices R&D, Helsinki and Petteri Lundén from Nokia Research Center, Helsinki in their study of VoIP on E-UTRA.

## REFERENCES

1. Holma H, Kuusela M, Malkamäki E, Ranto-aho K, Chen T: VoIP over HSPA with 3 GPP Release 7. *Proceedings of IEEE International Symposium on Personal, Indoor and Mobile Radio Communications, September 2006, Helsinki, Finland*

2. Wang B, Pedersen KI, Kolding TE, Mogensen PE: Performance of VoIP on HSDPA. *Proceedings of the 61st Vehicular Technology Conference (VTC '05), June 2005, Stockholm, Sweden* 2335-2339.

3. Lundén P, Kuusela M: Enhancing performance of VoIP over HSDPA. *Proceedings of the 65th Vehicular Technology Conference (VTC '07), April 2007, Dublin, Ireland* 825-829.

4. Chen T, Kuusela M, Malkamäki E: Uplink capacity of VoIP on HSUPA. *Proceedings of the 63rd Vehicular Technology Conference (VTC '06), July 2006, Sydney, Australia* 451-455.

5. Third Generation Partnership Project : Evolved Universal Terrestrial Radio Access (E-UTRA); Medium Access Control (MAC) protocol specification. TS 36.321, Release 8, 2008

6. Love R, Kuchibhotla R, Ghosh A, Ratasuk R, Xiao W, Classon B, Blankenship Y: Downlink control channel design for 3GPP LTE. *Proceedings of IEEE Wireless Communications and Networking Conference (WCNC '08), April 2008, Las Vegas, Nev, USA* 813-818.

7. Nory R, Kuchibhotla R, Love R, Yakun S, Weimin X: Uplink VoIP support for 3GPP EUTRA. *Proceedings of the 65th Vehicular Technology Conference (VTC '07), April 2007, Dublin, Ireland* 710-714.

8. Persson F: Voice over IP realized for the 3GPP long term evolution. *Proceedings of the 66th Vehicular Technology Conference (VTC '07), October 2007, Baltimore, Md, USA* 1436-1440.

9. Fan Y, Kuusela M, Lundén P, Valkama M: Downlink VoIP support for Evolved UTRA. *Proceedings of IEEE Wireless Communications and Networking Conference (WCNC '08), April 2008, Las Vegas, Nev, USA* 1933-1938.

10. Fan Y, Lundén P, Kuusela M, Valkama M: Efficient semi-persistent scheduling for VoIP on EUTRA downlink. *Proceedings of the 68th Semi-Annual IEEE Vehicular Technology (VTC '08), September 2008, Calgary, Canada*

11. Puttonen J, Kolehmainen N, Henttonen T, Moisio M: Persistent packet scheduling performance for voice-over-IP in evolved utran downlink. *Proceedings of the 19th IEEE International Symposium on Personal, Indoor and Mobile Radio Communications (PIMRC '08), September 2008, Cannes, France*

12. Rinne M, Kuusela M, Tuomaala E, Kinnunen P, Kovács I, Pajukoski K, Ojala J: A performance summary of the evolved 3G (E-UTRA) for voice over internet and best effort traffic. *IEEE Transactions on Vehicular Technology* 2009, 58(7):3661-3673.

13. Puttonen J, Puupponen HH, Aho K, Henttonen T, Moisio M: Impact of control channel limitations on the LTE VoIP capacity. *Proceedings of the 9th International Conference on Networks (ICN '10), April 2010, Menuires, France* 77-82.

14. Holma H, Kallio J, Kuusela M, *et al.*: Voice over IP (VoIP). In *LTE for UMTS: OFDMA and SC-FDMA Based Radio Access*. Edited by:

Holma H, Toskala A. John Wiley & Sons, New York, NY, USA; 2009:259-281.

15. López Villa D, Úbeda Castellanos C, Kovács IZ, Frederiksen F, Pedersen KI: Performance of downlink UTRAN LTE under control channel constraints. *Proceedings of the 67th Vehicular Technology Conference-Spring (VTC '08), May 2008, Singapore* 2512-2516.

16. International Telecommunication Union, Rep. ITU-R M.2134: Requirements related to technical performance for IMT-Advanced radio interface(s). 2008

17. Third Generation Partnership Project: Physical Layer Aspects for evolved Universal Terrestrial Radio Access (UTRA). TR 25.814, Release 7. version 7.1.0, September 2006

18. International Telecommunication Union: One way transmission time. Recommendation ITU-T G.114, February 2003

19. Orange, China Mobile, KPN, NTT DoCoMo, Sprint, T-Mobile, Vodafone, Telecom Italia : LTE physical layer framework for performance verification. R1-070674, 3GPP TSG-RAN1#48, St. Louis, Mich, February, 2007

20. Fan Y, Lundén P, Kuusela M, Valkama M: Performance of VoIP on EUTRA downlink with limited channel feedback.*Proceedings of IEEE International Symposium on Wireless Communication Systems (ISWCS '08), October 2008, Rekjevik, Iceland* 79-84.

21. Sampath A, Kumar PS, Holtzman JM: On setting reverse link target SIR in a CDMA system. *Proceedings of the 1997 47th IEEE Vehicular Technology Conference. Part 1 (of 3), May 1997* 929-932.

22. Jalali A, Padovani R, Pankaj R: Data throughput of CDMA-HDR a high efficiency-high data rate personal communication wireless

system. *Proceedings of the 51st Vehicular Technology Conference (VTC '00), May 2000, Tokyo, Japan* 3: 1854-1858.

23. Brueninghaus K, Astély D, Sälzer T, Visuri S, Alexiou A, Karger S, Seraji G-A: Link performance models for system level simulations of broadband radio access systems. *Proceedings of IEEE International Symposium on Personal, Indoor and Mobile Radio Communications (PIMRC '05), September 2005, Berlin, Germany* 4: 2306-2311.

24. Third Generation Partnership Project: Mandatory speech CODEC speech processing functions; AMR speech Codec; General description. TS 26.071, July 2007

25. Internet Engineering Task Force: Robust Header Compression (ROHC). IETF RFC 3095, July 2001

26. International Telecommunication Union: Guidelines for evaluation of radio interface technologies for IMT-Advanced. Rep. ITU-R M.2135, 2008

27. Ericsson: Radio characteristics of the ITU test environments and deployment scenarios. R1-091320, 3GPP TSG-RAN1#56bis, Seoul, Korea, March 2009

9

# SOLUTIONS COMPARISON TOWARDS VOICE SERVICES IMPLEMENTATION FOR OPERATORS STARTING LTE DEPLOYMENT

**Youness Jouihri, Zouhair Guennoun**

Laboratoire d' Electronique et de Communications—LEC, Ecole Mohammadia d' Ingénieurs—EMI, Université Mohammed V-Agdal—UM5A, Rabat, Morocco

## ABSTRACT

The continuity of voice services, for many operators planning to start their LTE network, is one of the key requirements to cover back the cost of this expensive network. In this paper, we discuss and compare the possible solutions to add voice services on an LTE network dedicated mainly for high-speed data transfer. By comparing advantages and inconvenient of the existing approaches, we conclude that the selection of the One Voice approach is obvious, especially for operators starting LTE deployments, the One Voice approach has the support of all the famous operators and vendors. Selecting the One Voice approach as a first alternative was evident, but the choice of CSFB as a mid-term solution or backup solution from many operators was not clear. As a key lesson can be learned from the existing technologies, CDMA and GSM, where the type of handset was one of the major factors of the GSM success; we prove in this paper that VOLGA approach is better than CSFB for many reasons: The VOLGA approach uses less complex and less expensive handset while it gives more comfortable user experience compared to CSFB approach, so operators should take into consideration these two points before selecting the second alternative.

## KEYWORDS

Voice over LTE; One Voice; VOLGA; IMS; SIP; CSFB; SRVCC

## 1. INTRODUCTION

LTE is a full IP network designed for data transfer. Moving towards this new technology, mobile operators who receive "over 80% of their revenues from Voice and SMS" [1], can't ignore the continuity of these two services over data services.

The LTE will be more attractive for operators, if it can replace a hundred years of developments in voice technologies, and provide the same quality or maybe better.

However, until now, voice format is not "defined as a major omission" [1] for the LTE system. Therefore, operators and vendors have to agree on the best solution to transfer the voice services, in this all-IP world.

The remaining sections of this paper are outlined as follows: After an introduction to the SIP protocol, its benefits for the operator and its role to enable the voice over IP services in section 2. Section 3, explains why IMS is the preferred choice for operators to provide Voice over LTE and its biggest limitation. Section 4 is reserved to present Interim mechanisms used to make the transition to IMS-based call control. Section 5 introduces the VOLGA approach and section 6 is devoted to the One Voice approach, its advantages and comparison with VOLGA approach. Finally, a summary is provided in Section 7.

## 2. SIP PROTOCOL, EVOLUTION, ROLE AND BENEFITS

The tendency that all communications are moving to an IP based world was obvious since last few years. The standard to use on this all-IP world was a challenge for mobile operators. SIP protocol was selected and standardized to enable those services related to Voice over IP [2]. **Figure 1** is an example where SIP connects the circuit switch to IP Multimedia Subsystem.

For many operators, especially those using MSC Server, the SIP protocol implementation can be very easy, as it is a simple feature working with other MSC Server (MSS) features, and implements the Media Gateway Control Function (MGCF) in the MSS [3]. With the MGCF, the IMS and CS can easily communicate by connecting it to GMSCS (Gateway MSC Server), this gives the networks the CS Interworking functionality with IMS, and so, the VOIP and Video can be forwarded to CS mobile user. The functionality of SIP can be extended to different interfaces, if we can add the ISUP tunneling facility, for that both SIP-I (Internet Engineering Task Force (IETF)) with ISUP tunneling and SIP-T (International Telecommunications Union (ITU-T) without tunneling should be defined [3].

Towards the IMS, only SIP-T is used without ISUP tunneling. Session refreshment messages are supported on all the interfaces [3]. SIP-T protocol is simply a SIP-I protocol with tunneling OFF [3].

This separation based on ISUP tunneling facility gives a major benefit for this protocol, it can connect the Circuit-Switched (CS) and the IP Multimedia Subsystem (IMS) domains, and provide an open ISUP tunneling SIP trunk (soft switch) interface (SIP-I), that can be used either between the MSSs or towards other equipment's having SIP-I interface, like VMSs (Voice Mail Systems), ATA (Analogue Telephone Adapters), SIP-I gateways and so on.

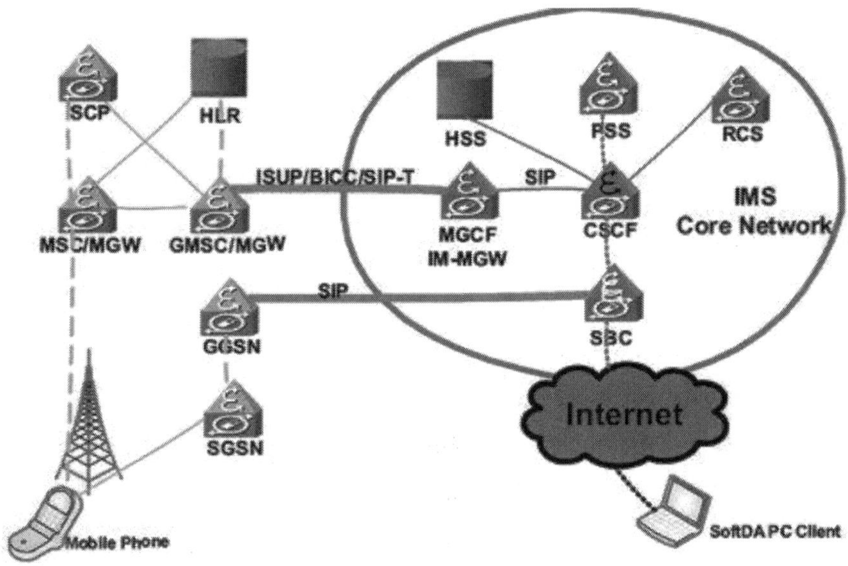


**Figure 1**. SIP connects the circuit-switched and the IP Multimedia Subsystem domaine (more clarifications about components and interfaces given in [4]).

SIP-I is preferred between MSSs, because it supports GSM or ISDN services more effectively [4]. This feature complies with the 3GPP IMS standards and with the relevant IETF standards. Both the User Datagram Protocol (UDP) and the Transmission Control Protocol (TCP) are supported as signaling transport for SIP. In addition, semi-permanent Stream Control Transmission Protocol (SCTP) associations can be used optionally to transport SIP messages. With SCTP transport, multi-homing can be used as well [4].

# 3. IMS ARCHITECTURE, EVOLUTION AND LIMITATION

The main focus for Operators starting LTE deployment is the way of delivering their main catch cow "Voice" over LTE, as no clear strategy was defined [1]. Several propositions have been suggested, but IMS was always the favorite choice. With its sophisticated architecture and feature set, developed after more than ten years work by 3GPP, and thousands of pages in different specifications, which in a "way made sure of the continuity of service, interoperability and roaming between operators". Also it can lead the voice evolution story from existing CS networks to a word of IP, after the new convergence to a packet-based access and core network [5]. The IMS can easily connect the Circuit switch and packet switch domain to the applications and services domain, using the facilities given by its main Components shown on Figure 2.

The IMS evolution was very fast, see **Figure 3**, from one version to the next, more services was provided for operators; but most of the services related to VOIP and CS interworking, subject of this paper was defined on Release 8.

The IMS release 8 was a smooth evolution from previous release building new capabilities:

- IMS centralized services: allow devices using CS connection (GSM/3G CS radio) to use IMS services.
- Multimedia session continuity: improves voice call continuity feature to enable continuity of multimedia media streams when IP access is changed.
- Single-Radio Voice Call Continuity: enables seamless voice handover from LTE (IMS VoIP) to CS.

From the above description we can believe that IMS is the best option for Voice over LTE. IMS release 8 is a complete solution for transferring voice traffic on this all IP word designed for data. With its golden list of features set, we can overcome all the problems and limitations related to voice integration, on a network designed for data.

More than giving smooth continuity of the most important service for operators, IMS is the glue between different domains, so "legacy networks won't go away, IMS supports co-existence with voice infrastructure", and the handover to a 2G or 3G circuit switched channel can be done easily once running out of LTE coverage [4].

In parallel IMS support multitude of Application Servers that may give multiple functionality, and realize all the multimedia dreams for operators, but all these facilities make IMS implementation difficult. IMS, unfortunately, is suffering from its own complexity, having a sophisticated but heavy infrastructure, supporting many feature set developed after years of work, the full IMS implementation will cost a lot of time and money for operators, without having any guarantee that end-user will really need and go for all these new services, and so, no insurance to cover back the cost of this big investment. This is why; operators should not go for the IMS bigbong approach.

For Operators starting LTE deployment, the complexity and the multimedia dreams have to be thrown away and they should concentrate on the voice service only, which is the real reason for introducing IMS over their LTE network, "an impossible suggestion only a short time ago for all who select IMS approach, but the pressure seems to be mounting" [6], they cannot concentrate on implementing a full IMS network and leave their main objective for creating a real network IP based. The first focus should be on voice, make it working, instead of investing heavily in building the infrastructure and waiting for the subscribers and services to come, after that they can target the services that subscribers are likely to embrace then, building up the infrastructure to support those services incrementally, by adding the IMS-peripheral components needed, also check the possibility to reuse smartly their existing network, by using facilities that can change their components to IMS component as shown in session 2, where operators using MSC servers can implement the Media Gateway Control Function (MGCF) in their MSS by adding SIP feature. The replacement of the entire functional systems will not be required, till reaching a point where the sophistication of the services and the operational saving of IMS will become obvious; after that the adoption will accelerate toward a full IMS network.

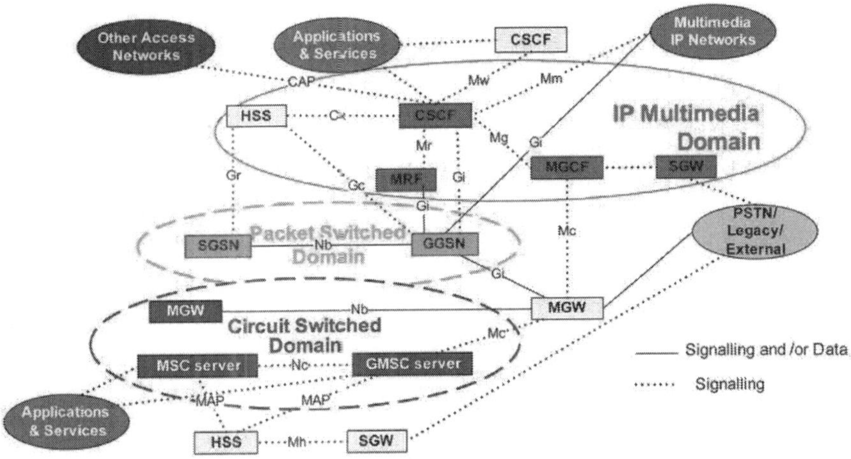


**Figure 2**. IMS detailed architecture (Release 5) (Interfaces description and components given in [4]).

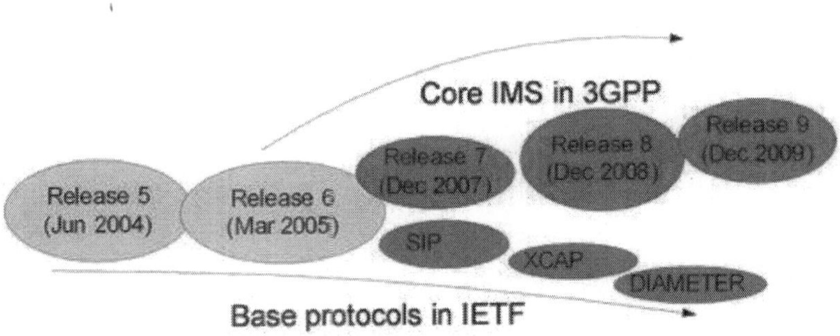


**Figure 3**. Evolution of IMS in relationship with SIP protocol [4].

The IMS pragmatic approach is needed as the full implementation can take several years. So, which Interim mechanisms can be selected to make the transition happen?

# 4. INTERIM MECHANISMS FOR IMS-BASED CALL CONTROL TRANSITION

Today, all the parties agreed, 3GPP and GSMA, that IMS will become the mainstream choice for voice over LTE, and will be the standard architecture for mobile voice [7], but as explained above, operators are not yet ready to adopt this approach, even more, handset makers has to create devices that support 3GPP Release 8 and Release 9 functions that work with a VOLTE network. To give more time for network and handset to be ready to support IMS big-bong approach, the use of interim mechanisms became mandatory. Many approaches already exist, and the best selection will give definitively a smooth IMS deployment. So let us describe these approaches before making any selection.

The first approach, circuit switched fallback (CSFB), where LTE will provide just data services, and for the initiation or the reception of a voice call, fallback to CS domain will be performed [7].

Figure 4 illustrates the case, when a subscriber wishes to make a voice call, in a CS Fallback to legacy 3GPP network architecture, the UE makes a service request to the LTE network, and the coordination with the 3GPP network will redirect the UE to the legacy CS network. For mobile terminated calls, the subscriber is paged in the LTE network and he cannot move to CS network till subscriber decides to accept the call [8]. For SMS, it can be sent without leaving the LTE network (the handset can use interface known as SGs which allows messages to be sent over an LTE channel [1]).

For many operators, they need just to upgrade their MSC to support CSFB [7].

However, CSFB may cause LTE data connection drop and entail an additional call set up delay [7]. The CSFB might not be the best interim solution, but one of the most important addition that we can get by adopting it, is the scenario of LTE roaming [7]. "For example, when the visited network does not have the IMS or when the IMS roaming protocol is yet to be deployed, CSFB can provide voice-call service for inbound LTE roamers" [7].

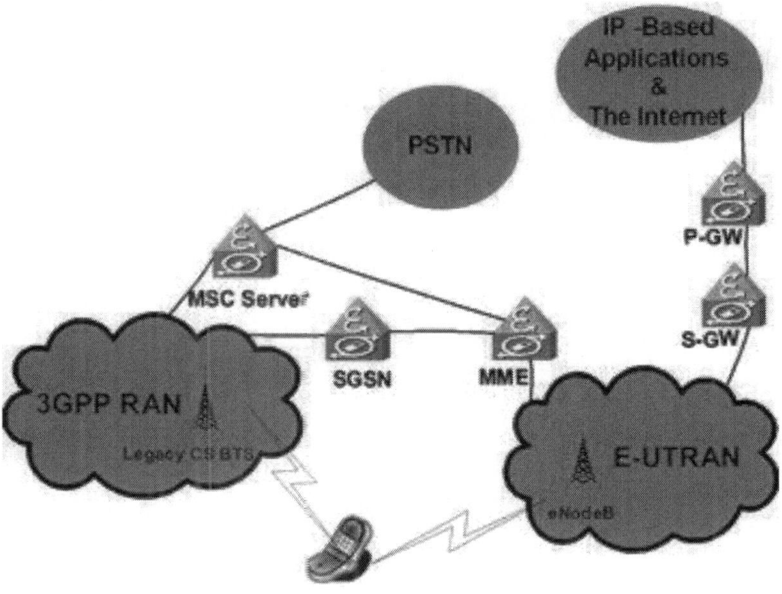


**Figure 4**. Circuit switched fallback architecture [8].

The second approach, dual radio handset approach or simultaneous voice and LTE (SVLTE), is "solely based on the handset", no changes are required on the network and the IMS deployment is either not needed, the handset will simultaneously get the data services from the LTE and the voice service from CS mode. This solution can suffer from the fact that "the phone can become expensive with high power consumption" [7].

The third approach is the approach founded on the usage of over-the-top (OTT) where applications like Skype and Google Talk can provide LTE voice service. The LTE "can create natural convenience for the development of OTT voice calls" and break almost all the barriers, by using its features like broad bandwidth, low latency, being always-online, and All-IP [7].

This approach cannot receive much support, as no operator can handover completely its main driver of revenue, voice, to the OTT actors.

For the fourth and fifth approaches, VOLGA and One Voice are the most applicable approaches to meet the consumers' expectations, according to their impressive list of supporters. Very big debate and polemics were initiated on different forums to decide which approach will be the final solution to exceed the voice challenge on LTE network. Next two chapters will be reserved for these two approaches since more details and clarifications are needed to reach a verdict about the best option to be adopted.

# 5. THE VOLGA (VOICE OVER LTE VIA GENERIC ACCESS) APPROACH

The VOLGA approach was initiated during the VOLGA Forum on March, 2009 for extending traditional voice and SMS services over LTE access networks, without requiring a re-architecture of the network, one of the very low cost methods of getting the two highest drivers of revenue services for operators [1]. The concept is very simple, instead of creating the core telephony services in IMS, we simply make the existing telephony infrastructure a packet service delivered over IP via LTE, which means making use of GSM signaling over IP, and elevate the core voice network to be able to deliver packet service over the LTE access network. The main idea was to adapt Universal Mobile Access/Generic Access Network (UMA/GAN) system to LTE; "Whether through Wi-Fi radios (Nanoradio, Swedish Wi-Fi radio start-up) or fixedline services (T-Mobile's highly successful fixed-line service), it has been proven that the UMA/GAN is a very successful platform to deliver mobile voice (and data) services over IP access networks" [9], so we should also check it on the LTE networks.

The aim of VOLGA is to make use of the existing mobile core, which is extremely feature rich, to deliver the primary revenue generating service for operator, by "tunneling circuit switched voice traffic across LTE from the generic access network controller (GANC) server" to the evolved packet core in LTE [9], as shown in **Figure 5**. For handover from LTE to 2G/3G CS, the feature Single Radio Voice Call Continuity (SRVCC) can be used [10]. The use of this feature contributes a lot to make VOLGA better option, in comparison to the previous interim approaches, especially CSFB. In addition of enabling the full range of the current voice services and made them available, contrarily to the services given by CSFB, the SRVCC gives subscribers chance to use a single radio mode device, which is less complex and less expensive, also the switching to CS network will be needed only when terminal roams out of LTE coverage, so the call setup time will be less, since time is required only out of LTE [11]. On the other side CSFB needs a dual mode device, complex and expensive due to the increase of network signaling load, also the switching to CS network is needed for every mobile originating or terminating call, which augment the voice call setup [11].

So from the above description, VOLGA beat CSFB with the various advantages mentioned previously, we can even add more, like the protections of the existing and expansive voice core investment, the smart exploitation of an existing standard GAN already "proven as an effective and robust method for extending mobile voice services over broadband packet access networks" [9], in addition VOLGA is supported by all major handset vendors and it can clear all the confusion for an early handset development [9]. This confusion can be cleared by "defining the LTE voice to be identical to the existing voice services of the 3G and 2G networks" [9], so LTE will not face the same problem of the 3G's deployment where "operators waited years for viable handsets" [9].

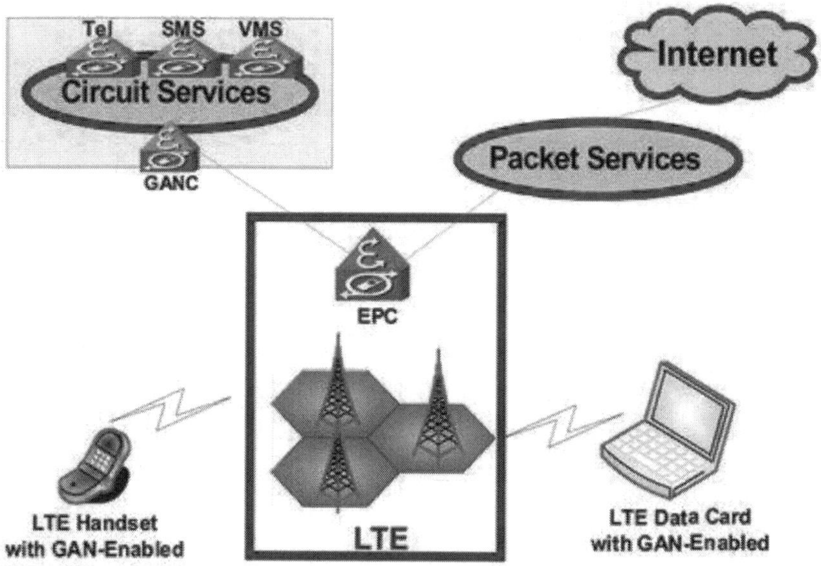


**Figure 5**. Elevate existing core voice services to become packet services over LTE [9].

The VOLGA can also win the battle hands down against the dual radio handset approach by referring to the type of handset that can be used by the last approach, expensive and should support the high power consumption. A key lesson can be learned from the existing technologies, CDMA and GSM. Many things makes GSM more famous than CMDA, but I think, the major factor of success was the handset, GSM gave more flexibilities (change the phone, buy new versions) and choices to the end user than CDMA.

For the OTT approach, there is no need to go through details, and make technical comparison between this approach and the VOLGA. Logically no operator can give the best source of revenue to OTT actors, so the VOLGA selection will be obvious, since it will keep their resource save.

The VOLGA can be considered as a mid-term solution for operators preferring the migration of all services to IMS; as we proved previously that this approach has more chance than all previous approaches, to replace IMS for telephony services delivery [9]. The VOLGA can also be considered as a long term solution for operators willing to invest in LTE for non-voice services, and keep their existing 2G/3G network for the telephony services [9]. The handset and the telephony services must support 2G/3G as a fall back even after full LTE deployment, this can guaranty the use of VOLGA as long term proposition [9]. The VOLGA solution is not yet standardized by 3GPP, many operators and vendors were supporting this approach, but now they join the group adopting the new approach called One Voice [12]. So what is this approach? And how it took the support of many major vendors from VOLGA?

## 6. THE ONE VOICE APPROACH: ITS ADVANTAGES AND ITS COMPARISON WITH VOLGA APPROACH

The One Voice approach was announced on November 2009 after coordination between "six operators, three core networks vendors and three handsets manufacturers". The idea was to use the concept of IMS MMTel feature, already defined by 3GPP in IMS release 6, "to a minimum set of functionality for a voice service equivalent to what we have today in CS domain" [13]. After the adoption of this approach by GSMA on January 2010, One Voice started to attract different parties, especially when it was announced publicly on February 2010 at MWC (mobile word congress) [13]. GSMA has added three key interfaces, to address the entire end-to-end voice and SMS ecosystem, by also focusing on Roaming and Interconnect interfaces, in addition to the interface between customer and network [13].

The First interface, UNI (User Network Interface), as shown in **Figure 6**, is located between the user's and the operator's network. Define a common User Network Interface enables the benefice of maintaining roaming capability, giving Global scale and global ecosystem and greater amortization of R&D as a result of lower cost per unit for vendors [14].

The second interface, I-NNI (Interconnect Network Network Interface), as shown in **Figure 6**, is the interface located between the networks of two parties making a call, creating an end-to-end call connectivity, which required a common functionality and a common implementation to remove the need of expensive interworking functionality [14].

The third interface, Roaming Network Network Interface is an interface located between the home and visited network, used by users roaming in visited networks, not attached to their home network. It is the most challenging interface, as it required compatibility between different networks, and the incompatibility with any roaming solution of other networks, might potentially affect the voice service of the roamers on these networks [14].

The biggest challenge was the definition of a functional split between Home and Visited networks, as the 3GPP has different options for split [15], so the One Voice approach will select one. The selection will be based on the minimum requirements of this approach, means having the voice service equivalent to what we have today in CS domain.

The **Figure 7** shows the three roaming options given by 3GPP: The first alternative has the ability to meet the regulatory requirements for roamers in the visited network, such as Lawful intercept and Emergency call by using the VOIP emergency calls via visited network, also the LI is possible on the visited network [15].

**Figure 6**. User Network Interface (UNI) and Interconnect Network Network Interface (I-NNI) [14].

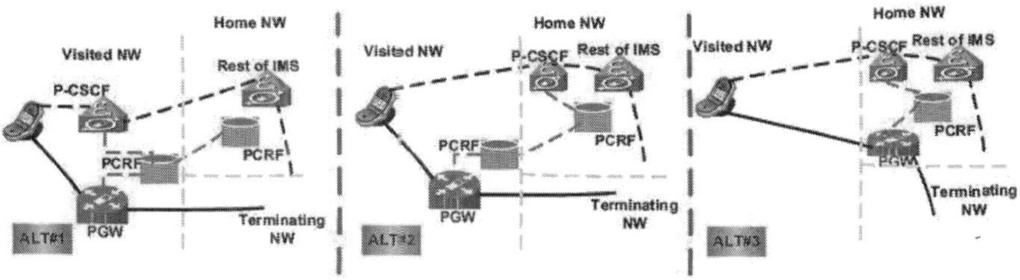


**Figure 7**. 3GPP architecture contains three roaming options [15].

The two other alternatives cannot give this ability; as for emergency call, the VOIP requires access via visited P-CSCF or emergency call via CS domain, and for Lawful interception, the Control plane interception is not possible if IPSec applied.

Obviously, alternative one will be selected, as it can provide the needed requirement. The functional split selected by One Voice makes the user experience more comfortable [15], as it gives the ability to optimize the media path for voice sessions, and minimize the number of media gateways and transcoding for user plan, which reduce the call setup time. After this selection we have a common roaming network interface which allow interoperator communication, policy control and enforcement, and inter operator charging.

The Work of GSMA will not be limited on the addition of new interfaces, but it will encompass the Key technical requirements to make a nice voice experience on a network having different technologies. For an operator starting LTE deployment, the LTE coverage will be very limit and handover to legacy networks will be a big requirement. So it will be necessary to ensure the continuity of the voice calls when user moves from LTE to legacy networks, this handover can be achieved using Single Radio Voice Call Continuity (SRVCC) [11]. The SRVCC is an optional feature included in One Voice specification, allows devices with one radio access technology at a time with an ongoing voice call, to transition to the circuit-switch domain in the event of loss of LTE coverage, this feature will significantly increase the complexity of the network as PS and CS domains need to support it [11], but for operators on the starting phase of LTE deployment, the selection of SRVCC will make their network more comfortable for users, because without it, they will suffer from the high number of calls drop related to the loss of LTE coverage. The complete handover procedure is given in [11]. For the SMS part, which is a CS service that needs special attention with regard to LTE, the Target SMS solution will be based on IMS when the UE and network may support the SMS-over-IP, the migratory solution will be SMS over SGs as defined in 3GPP [16].

As described before, the One Voice approach is an IP Multimedia Subsystem used at its minimum capacity to provide Voice over LTE, so the One Voice approach is definitively the best solution to reach the target solution IMS, and realize the multimedia dream of many operators, as we can provide any new service, that subscribers are likely to embrace, by adding the needed IMS-peripheral components, without replacing the entire functional systems, progressively we will reach the full IMS configuration. The VOLGA approach will use different system, so the pragmatic IMS approach cannot be adopted on it.

Today, the OTT actor's services became more sophisticated and less expensive, especially for international calls, so many users start using it in state of using expensive legacy network, and so, operators start losing some of their permanent source of revenue [17]. Providing the same voice experience as legacy network cannot be a permanent solution, as the competition is increasing day by day, and customers start adopting the new broadband wireless technology, so they will not accept lower quality voice service with higher price, operators should either give a very good billing discount which will affect their revenue, or either provide an HD (high definition) voice services, which enhances voice quality using a wider frequency range, so mobile operators should be able to offer nice voice quality, even higher than landline voice, like that they can compete the OTT actors and save their revenue.

VOLGA cannot give a voice quality better than the existing network, as the approach is based on the existing system, so the HD option is not possible on this approach.

Finally, the list of supporter of One Voice is still increasing, and the major operators and vendors are already there, some of them drop their support to VOLGA and join the One Voice group [12]. Having a big agreement for a specific technology makes life easy for operators especially in case of roaming, so the technologies collecting more support will be obviously the successful one and the best choice for any operator on the starting phase of deployment.

## 7. SUMMARY

This paper has analyzed the different solutions to implement the voice services over LTE network. After comparing these approaches, we have presented for operators starting their LTE network, the best selections to support the voice services over LTE. Our results match with the actual tendency, where all the major operators and vendors went for the One Voice approach as a primary alternative. But as a mid-term or secondary alternative, we proposed the VOLGA approach contrarily to a big number of operators who want to go for CSFB. As Yankee Group Research Director Brian Partridge said referring to VOLGA approach, "The reason you haven't seen a lot of operators is because they haven't studied it yet" [18]. On this paper we conclude, after studying all aspects, that we should not throw away the VOLGA approach. We proved that VOLGA is better than CSFB for many reasons. So, operators planning to start this big investment should go through all the details of VOLGA approach. As a wrong selection can badly affect the evolution of a new technology, and the history of mobile technologies have already proved it.

## ACKNOWLEDGEMENTS

The authors would like to acknowledge Mr. Faisal M. J. Attieh for his valuable advice and support of this work.

## REFERENCES

1. Cellular telecoms, "Voice over LTE-VoLTE," 2011. http://www.radio-electronics.com/info/cellulartelecomms/lte-long-term-evolution/voice-over-lte-volte.php

2. K. Banerjee, "SIP (Session Initiation Protocol) Introduction," 2009. www.siptutorial.net/SIP/index.html

3. Nokia Siemens Networks, "Feature 1331: Session Initiation Protocol in the MSC Server," 2009.

4. A. Bencheikh, "Introduction to IMS & SDP," Al Akhawayn University, Ifrane, 2010.

5. AT & T, et al., "One Voice: Voice over IMS profile," 2009. www.nokiasiemensnetworks.com/system/files/document/One_Voice_Profile_0.pdf

6. M. Sauter, "The IMS One Voice Profile—Some Thoughts," 2009. http://mobilesociety.typepad.com/mobile_life/2009/11/the-ims-one-voice-profile-some-thoughts.html

7. Q. h. Chen, "Evolution and Deployment of VoLTE," 2011. www.huawei.com/en/static/hw-094164.pdf

8. I. Tanaka, T. Koshimizu and K. Nishida, "CS Fallback Function for Combined LTE and 3G Circuit Switched Services," NTT DOCOMO Technical Journal, vol. 11, no. 3, 2009, pp. 13-19.

9. S. Shaw, "LTE: Delivering Telephony Services in an LTE Network," 2009. www.smart-wi-fi.com/pdf/uma_mag_spring_09_lo.pdf

10. Ericsson White Paper, "voice over LTE," 2010. www.ericsson.com/res/docs/whitepapers/voice-over-lte.pdf

11. S. Vittal, "SRVCC: Single Radio Voice Call Continuity with LTE," 2011. www.slideshare.net/allabout4g/wpsrvccwithltepdf

12. M. Donegan, "MWC 2011: Deutsche Telekom Ditches VoLGA," 2011. www.lightreading.com/document.asp?doc_id=204581

13. D. Warren, "Voice & SMS over LTE Update," 2011. https://infocentre.gsm.org

14. D. Warren, "Voice over LTE," 2010. https://infocentre.gsm.org

15. Nokia Siemens Networks, "GSMA LTE Voice roaming," 2010. https://infocentre.gsm.org

16. D. Warren, "Next Steps for VoLTE," 2010. https:// infocentre.gsm.org

17. Juniper Networks, "The Evolution of Business Services: Delivering Value to Boost Revenue," 2011. www.juniper.net/us/en/local/pdf/whitepapers/2000305-en.pdf

18. S. Shaw, "Let's Give 'Em Something to Talk About," 2011. http://voiceoverlte.typepad.com/voice-and-sms-over-lte/2009/10/lets-give-em-something-to-talk-about.html

10

# IMPROVING MANAGEMENT PERFORMANCE OF P2PSIP FOR MOBILE SENSING IN WIRELESS OVERLAYS


**Pablo Sendín-Raña [1], Francisco Javier González-Castaño [2,3], Felipe Gómez-Cuba [2], Rafael Asorey-Cacheda [1,*] and José María Pousada-Carballo [1]**

[1]Centro Universitario de la Defensa, University of Vigo, Praza de España, s/n, 36920 Marín, Spain

[2]AtlantTIC, University of Vigo, Campus s/n, 36310 Vigo, Spain

[3]Gradiant, Campus s/n, 36310 Vigo, Spain


## ABSTRACT


Future wireless communications are heading towards an all-Internet Protocol (all-IP) design, and will rely on the Session Initiation Protocol (SIP) to manage services, such as voice over IP (VoIP). The centralized architecture of traditional SIP has numerous disadvantages for mobile *ad hoc* services that may be possibly overcome by advanced peer-to-peer (P2P) technologies initially developed for the Internet. In the context of mobile sensing, P2PSIP protocols facilitate decentralized and fast communications with sensor-enabled terminals. Nevertheless, in order to make P2PSIP protocols feasible in mobile sensing networks, it is necessary to minimize overhead transmissions for signaling purposes, which reduces the battery lifetime. In this paper, we present a solution to improve the management of wireless overlay networks by defining an adaptive algorithm for the calculation of refresh time. The main advantage of the proposed algorithm is that it takes into account new parameters, such as the delay between nodes, and provides satisfactory performance and reliability levels at a much lower management overhead than previous approaches. The proposed solution can be applied to many structured P2P overlays or P2PSIP protocols. We evaluate it with Kademlia-based distributed hash tables (DHT) and dSIP

## KEYWORDS

P2P overlays; DHT; P2PSIP

## 1. INTRODUCTION

Wireless communications are heading towards an all-Internet Protocol (all-IP) approach, in which phone services provided by IP multimedia subsystems (IMS) will rely on the Session Initiation Protocol (SIP) for management purposes. However, traditional SIP has a centralized architecture, and therefore, peer-to-peer (P2P) networking may improve its resilience in wireless environments.

SIP is a signaling protocol commonly used for network management and data exchange applications, such as voice over IP (VoIP) or sensor information. Among other features, SIP provides high scalability and flexibility. These are desirable features in heterogeneous mobile networks of arbitrary size [1] in which people no longer only communicate in the traditional way, but also share other information that is relevant to the *context* of their activities [2]. In fact, nowadays, many phone services rely on context-aware information provided by *embedded sensors*, which the users share. This imposes the need to achieve fast and seamless data delivery between network devices, for which SIP offers reliable support. In the context of mobile sensing, P2PSIP protocols facilitate decentralized and fast communications between sensor-enabled terminals. Nevertheless, in order to make P2PSIP protocols feasible in mobile sensing networks, it is necessary to minimize overhead transmissions for management purposes, thereby reducing battery lifetime.

P2P services are often based on overlay networks, in which resource management is a key challenge. P2P network states are strongly transient, and routing survivability to failures is a very important design consideration [3]. In practice, churn (peers leaving or joining the network) can affect the accuracy of the routing and resource tables at the peers, because certain entries (corresponding to individual peers) may be missing or stale. This affects both the efficiency and consistency of operations, such as lookups. While incomplete distributed hash tables (DHT) degrade performance (the less complete the registration information, the more hops the lookup procedure needs), stale contacts increase latency, due to timeouts (when peers that have waited a defined time for a response decide that the requested peer no longer belongs to the overlay). Since timeout intervals typically last no longer than a few round-trip times, latency can easily exceed the needs for common real-time operations. This is a greater problem in mobile wireless networks, where topology is inherently more unstable than in the wired Internet.

Peers that participate in a P2PSIP [4] overlay not only act as conventional SIP user agents (UAs), but also, collectively, play the normal roles of a central server. In such a system, registrar servers (where registration information is placed), proxy servers (intermediary nodes) and message routing functions of the traditional SIP protocol are replaced by a distributed P2P overlay. The core functions of this overlay are storage, discovery and access to digital resources. Currently, the P2PSIP working group of the

Internet Engineering Task Force (IETF) is considering proposals for the P2PSIP peer protocol based on distributed SIP (dSIP) [4] and Kademlia [5]. dSIP is an SIP-based protocol that has been proposed as a generic framework for a distributed SIP location service. In other words, it is a protocol for resource lookup in peer-to-peer SIP networks. It is simple and reusable with traditional SIP UAs. We describe Kademlia in Section 2.3.

P2PSIP with DHT is a realistic alternative for large decentralized deployments [6]. According to a study of three different evolution paths for mobile peer-to-peer (MP2P) communications in [7], it is a strong candidate for Internet VoIP support in worldwide interoperability for microwave access (WiMAX) or long term evolution (LTE) networks. Kademlia and dSIP are suitable for P2PSIP implementation on wireless networks. In this context, they yield good routing performance and scalability, allow self-organization and achieve secured, trustworthy P2P overlays. In [8,9], it is demonstrated that a Kademlia-based implementation can support general Internet services and distributed multimedia services on wireless networks. Kademlia's exclusive-OR (XOR) topology-based routing schema requires less maintenance than other structured P2P overlays [10], such as Pastry [11] or Tapestry [12]. It also performs better than Chord [13] in terms of lookup routing, overlay routing scalability and overlay self-organization.

From a practical perspective, we chose Kademlia for our study, because there is a real implementation of dSIP + Kademlia available (from the University of Parma [14]); the dSIP implementation is fully compatible with current commercial SIP solutions, and it is relatively simple to modify without compromising compatibility.

Some goals of P2PSIP systems are:

- Automatic setup: neighbor discovery or initial registration should be automatic procedures.

- Efficient lookup: the system should scale well with an increasing number of peers and growing demands.

- Support for heterogeneous peers: the participants in the overlay should be able to own different resources and have different network and available capacities; the system must also be platform-independent.

- Interoperability: A P2PSIP and another SIP-compliant node (even with traditional SIP implementations) should understand each other.

In order to keep the DHT registers of the P2PSIP network up-to-date, they are refreshed periodically. This paper discusses previous algorithms for managing *time-to-refresh (TTR)* timers and proposes an adaptive algorithm to optimize these timers to ensure the consistency of distributed routing tables and resource registers in P2P wireless overlay networks. This goal is important, since it will improve the trade-off between management cost and updating accuracy. With static *TTR* algorithms, management load grows with updating accuracy. The proposal, called adaptive *TTR* (*ATTR*), outperforms previous approaches in terms of management signaling overhead, by taking into account latencies between wireless network peers and by setting up a separate *TTR*timer for each P2PSIP resource to be refreshed, rather than a common timer per peer. Our work focuses on structured overlay networks [15], such as Kademlia [16] (employed by the eDonkey P2P system [17,18]) and Chord [13]. As previously mentioned, a joint Kademlia-based DHT/dSIP implementation of P2PSIP was considered. Following the Kademlia philosophy of favoring stable nodes, the algorithm adjusts *TTR* values depending on the permanency of the nodes in the overlay network (the longer the permanencies, the higher the values).

For the sake of clarity, we reproduce here some definitions in overlay terminology:

- Graceful and ungraceful leaving: A graceful peer departure from the overlay implies a notification to the overlay in which the peer transfers its routing and resource information. In an ungraceful departure (*i.e.*, due to link failure), the registered information on the lost peer stays in the overlay.

- Parallelism ($\alpha$): the number of parallel messages that a peer can send to other peers to accelerate operations.

- Replication ($r$): the number of peers responsible for each contact or resource information.

- System-wide number ($k$): the number of entries in each list or bucket of peers composing the DHT.

The rest of this paper is organized as follows. Section 2 describes the background. Section 3 discusses related work. Section 4 presents our proposal for adaptive network updates. Sections 5 and 6 evaluate the proposal by means of analysis techniques and simulations, respectively. Finally, section 7 concludes the paper.

# 2. BACKGROUND

## 2.1. SIP and Sensor Information Exchange

SIP was originally designed for VoIP and instant messaging applications. However, it is currently being used in applications beyond its original purpose, such as sensor information exchange. For example, SIP has been proposed as a protocol for metering information exchange in the Smart Grid [19,20], intelligent transport and mobility systems [21–23] and mobile service provision [24]. All these services rely on information provided by sensors.

Traditionally, wireless operators have largely ignored device-to-device (D2D) communications, because they considered that they were only useful to reduce the cost of service provision [25,26]. However, in recent years, this has changed, mainly for two reasons. Firstly, practical context-aware applications have emerged thanks to the many sensors available in current commercial terminals and their ubiquitous data connectivity. Since collaborative context-aware applications need to gather information from nearby devices, efficient D2D communications are necessary to discover these devices and exchange sensor information with them. Secondly, machine-to-machine (M2M) applications are growing fast. In this scenario, operators have already announced the development of the LTE-direct specification, for which P2PSIP may fulfill the requirements for information exchange between neighbor devices [27,28].

The popularization of mobile devices, such as smart phones, tablets and laptops, and the fast growth of wireless sensor networks, which are strongly power limited [29,30], have driven the development of energy-aware protocols. To avoid battery depletion, these protocols seek to minimize management message transmission. In this regard, P2PSIP is highly reliable and is also efficient in terms of power consumption [31].

## 2.2. P2P Overlays

P2P overlay networks are collections of nodes (called peers) interconnected by *logical* links (each logical link may comprise several physical links), conforming a distributed system without hierarchical organization or centralized control. Peers can join or leave the network at any time, meaning that typical overlay networks are dynamic and require some form of topology management.

The design space of P2P overlays is wide, and various classifications are possible. One of these divides P2P overlays into *structured* and *unstructured* overlays:

- Structured overlay. This involves key-based routing. The peers maintain routing tables and exchange these periodically for maintenance, *i.e.*, to support message delivery if there are no failures. Structured overlays are fully distributed and are usually characterized by a geometry, a routing algorithm and a maximum number of hops for routing a request. Since the information sought stays at specified locations, subsequent queries are efficient if the network is stable.

- Unstructured overlay. In this scenario, peers use flooding techniques to send queries across the overlay.

Flooding does not scale well. Moreover, some analytical studies on unstructured solutions often make topological assumptions that are incompatible with multi-hop wireless networks and ignore link vulnerability [32–34]. In structured overlay networks, the DHT mechanism allows self-organization with efficient routing, high search accuracy, high scalability and automatic load balancing [15]. We use the Kademlia DHT-based algorithm based on the XOR metric in our work. Nonetheless, our proposal can be extended to other structured overlays.

Structured overlay maintenance is costly, yet crucial, for P2P network performance. Most structured overlays use key-based routing, where each peer has a DHT to store a portion of the key-based routing information. DHT-based systems require $O$ (log $N$) hops on average to locate any data in the overlay.

Data-centric P2P applications commonly use replication to ensure high availability, even if the nodes leave the network ungracefully (without transferring information to other nodes). The peers exchange management messages periodically to keep published information in the overlay up-to-date. Frequent refresh of relevant data in the overlay results in high overhead, due to management messages. However, an overlay on top of a highly mobile wireless network is unreliable if refresh actions are too widely spaced.

Energy efficiency is another important issue in peer-to-peer overlay networks. Zhang *et al.* [35] demonstrated that transmitting one bit in a wireless network requires 1,000 times the energy of a single 32-bit computation. Obviously, this means that management techniques should use as few messages as possible.

## 2.3. Kademlia

Kademlia assigns 160-bit cryptographic hash IDs to peers and provides a lookup algorithm. Its main advantages are scalability and $O$ (log $n$) search complexity, $n$ being the number of peers. Each peer maintains a DHT with information about other peers to be used for routing schemas, and each Kademlia message includes peer-ID information about the sender.

Table 1 summarizes some notations used in the definition of Kademlia.

**Table 1.** Variable and function definitions.

| Symbol | Meaning |
|---|---|
| Peer-ID | Univocal peer identifier in the overlay. |
| Resource-ID | Univocal resource (user or service) identifier in the overlay. |
| $N$ | Size (in bits) of the peer-ID or resource-ID (Kademlia uses $N = 160$, the length of an SHA-1 digest). |
| $S$ | 160-bit identifiers for the space address. |
| $d$ | Distance function. |
| $p_i$ | Identifier of peer $i$. |
| $\oplus$ | Exclusive OR. |
| $N_{peers}$ | Number of peers in the overlay. |
| $N_{messages}$ | Number of management messages sent. |
| $N_{resources}$ | Number of resources that each peer registers in the overlay. |

Kademlia structures its ID space like a tree. The DHT is divided into *k*-buckets, where *k* is the number of entries in each bucket. A *k*-bucket is a list of *k* peers within a certain distance (in the XOR metric) of the current peer. For instance, the distance between 0001 and 0111 (four-bit IDs) in a XOR metric is six. This results in *kN* entries in the routing table.

Note that the XOR operation defines a non-Euclidean metric ($d(p_i,p_j) = p_i \oplus p_j$) with the following properties:

$$\begin{aligned} d(p_i, p_i) &= 0 \\ d(p_i, p_j) &> 0, \forall p_i \neq p_j \\ d(p_i, p_j) &= d(p_j, p_i), \forall p_i \neq p_j \\ d(p_i, p_j) + d(p_j, p_k) &\geq d(p_i, p_k) \end{aligned} \quad (1)$$

The XOR metric is unidirectional, which ensures that all lookups converge along the same path:

- Each *k*-bucket is sorted by last contact time (contact information, including peer-ID, IP address and port).
- When a message is received and the *k*-bucket is full, the receiver peer checks if the node in the *k*-bucket that was checked the last time is on-line: if there is no response, the node is removed, and the new peer is inserted at the tail. Otherwise, the new peer is discarded. Thus, Kademlia favors stable nodes.

The more contacts per bucket (achieved by increasing *k*), the more the DHT will be protected against churn. This is because peers are faster at repairing the routing table when a failure is detected. The DHT becomes more resistant to churn by accumulating high-quality contacts.

Kademlia uses strict parallel routing to accelerate registration and lookup operations according to the system-level concurrency parameter ($\alpha$). Increasing $\alpha$ reduces latency, but the overhead increases roughly linearly as a result [36].

Kademlia employs iterative, recursive or semi-recursive algorithms for node lookups. A lookup consists of a sequence of lookup steps (or hops). In iterative routing, an intermediate node always replies to the originating peer, which will then send the message to the peer that is the next hop. In recursive routing, an intermediate node in a path will send the messages directly to the next peer. The response unwinds and follows the same path back. Semi-recursive routing is similar to recursive routing, but the response is sent directly to the requester. In all cases, the initiating peer sends $\alpha$ messages to the closest nodes (to the requested key).

To ensure the persistence of <key, value> pairs, registry peers republish the keys after the *TTR* timer expires. Registrar peers, on the other hand, handle *time to expiration* (TTE) timers, which are typically initialized to twice the initial value of the *TTR*, so that on TTE expiration, the corresponding registered peer data is considered stale and removed. There is a timestamp associated with each contact or resource. If the timestamp corresponding to a particular contact or resource item has not been updated when the peers check their *k*-buckets and resource tables, the item will be considered stale and be removed (it is considered to correspond to a peer that has left the overlay ungracefully).

## 2.4. The P2PSIP Protocol

P2PSIP implements traditional proxy and registrar SIP functions in a distributed way. Resource information is distributed among all the peers in the overlay, and requests are also handled by the overlay infrastructure. The main advantages of P2PSIP are cost reduction and the elimination of single failure points. Using P2PSIP, any DHT-based P2P network can locate resources (services or users) in a decentralized way.

A P2PSIP deployment has an overlay name, and the participants can be peers (active) or SIP clients (passive). The peers are identified by a peer-ID, and they collectively serve as a directory service for locating resources using the defined DHT structure. The clients only use resources from the overlay. They do not participate in its maintenance.

The information stored in the P2PSIP overlay consists of registers of the peer nodes and the resources available at these nodes. Kademlia's distributed resource tables (DRTs) are similar to DHTs employed for peer registration. The same procedures as those described in Section 2.3, based on unique resource-IDs and a XOR metric, are employed. In order to keep registers up-to-date, they are periodically refreshed. Section 3 discusses previous algorithms for managing refresh time, whereas Section 4 describes the novel algorithm in the present work.

Protocol dSIP [5] uses SIP messages to implement P2PSIP, while preserving the semantics of conventional SIP messages as much as possible. Although there are newer P2PSIP implementations, dSIP has several advantages:

- Simplicity of implementation (still text-based)
- Minimization of the number of protocols required for a P2P UA.
- Easy integration into existing UAs and reuse of available SIP stack implementations.
- Widespread support.

The message to add, remove and query bindings in DHT and resource tables is SIP REGISTER [37]. dSIP supports:

- Peer and resource registration.
- Session establishment.
- DHT maintenance (dSIP is modular, so it allows multiple DHTs).

dSIP peers are active members of the overlay and provide operations to enable self-organization (SIP server-like functions) in addition to the basic functionality of any SIP endpoint. The dSIP overlay serves as a directory service for locating resources.

## 3. RELATED WORK

Li *et al.* compared the performance of the most relevant protocols for structured overlay networks under churn [38], including Kademlia, and concluded that they perform similarly with well-tuned periodical maintenance tasks. Maintenance of routing information stored in distributed DHTs is complex under churn, and it may saturate the network [39]. Although Li *et al.* focused on wired networks, their conclusions are also valid for wireless structured overlays. Cirani and Veltri defined a graceful leaving system, where peers issue unregistration messages for routing tables and resource registers [5]. In wireless networks, however, these messages may get lost, for instance, when a peer leaves the coverage area or loses connectivity due to terrain obstacles.

Successful implementations of structured overlays using Kademlia can be found in [40-43]. *K*-Umbrella [44] is a DHT algorithm that extends Kademlia using *k*-ary rather than binary trees.

Rhea *et al.* demonstrated by simulation that periodic recovery is more efficient than reactive recovery when a large overlay has reasonable churn rates [45]. Ou *et al.* evaluated Kademlia in mobile environments with different degrees of churn [8], setting adequate fixed refresh periods. They concluded that a choice of $k = 3$, $\alpha = 3$ and $r = 3$ makes the overlay resilient to ungraceful leavings. In [46], Koskela *et al.* proposed community overlays, an approach based on multiple overlapping overlays, and showed that they performed better. However, none of these analyses considered messaging load. On that topic, Maenpaa and Camarillo studied P2PSIP overlay network management in terms of signaling bandwidth, lookup delay and failed lookups for fixed refresh periods [47]. Chan *et al.* also measured bandwidth consumption for the same scenario [48]. Refresh periods for DHT algorithms were necessarily small for wireless networks under churn, implying significant signaling. Kelenyi and Nurminen reduced management traffic by means of probabilistic incoming message dropping [49], but this strategy affects routing performance.

Finally, some techniques employ a dynamic *TTR*. The schema in [50] follows a simple algorithm with additive increase and multiplicative decrease (*AIMD*). At each period, the algorithm defines a new *TTR* value based on the result of the previous updates. If resource locations do not change, the overlay is assumed to be stable, and the *TTR* can be incremented; otherwise, it must be decremented:

$$TTR = \begin{cases} TTR_{old} + C & \text{(nocongestion)} \\ TTR_{old} \times D & \text{(otherwise)} \end{cases} \quad (2)$$

where $C$, $C > 0$ and $D$, $0 < D < 1$ are constants. In addition, *AIMD* sets higher and lower limits for the new *TTR*values.

In the alternative approach in [51], refresh timers vary linearly. This approach therefore adapts poorly to network changes compared to *AIMD* in terms of the probability of stale information in the overlay. Accordingly, we selected *AIMD* to compete with our own dynamic proposal in Section 4.

Summing up, reducing the maintenance load of overlay networks over wireless networks is a challenging problem. The most advanced algorithms in the literature reduce this cost by dynamically adapting maintenance intervals (with dynamic refresh times). In this work, we present better *TTR* adaptation techniques with the same goal.

## 4. THE ATTR SCHEMA

Our schema seeks to provide adaptive *TTR* values to minimize management signaling load while keeping a low probability of unreliable information, which can degrade routing performance. As described in Section 2, Kademlia usually defines a fixed *TTR* value, although a dynamic value is technically possible.

In the original Kademlia implementation, and in the proposals in the literature, each peer has a single refresh timer. When the timer expires, the peer re-registers its information in the overlay.

In this work, we refer to a Kademlia overlay with static *TTR* values as the **original schema.**

Our *ATTR* solution involves different timers for each registered resource to populate peer information across the overlay (to update routing tables). Within the SIP REGISTER message, each peer sends a timestamp that the receiver uses to compute the corresponding latency. The response to this SIP REGISTER message includes that latency estimation, so that the sender can define the new timeout for the corresponding resource. The use of custom timers for each resource reduces message bursts considerably in comparison to the other proposals in our analysis.

As each peer in the overlay is responsible for a part of the DHT identifier space, peer churn causes two types of inaccuracies in routing and resource tables:

- Peers that still lack contact and resource information about a recently arrived peer.
- Peers with stale contact and resource information about recently departed peers.

The *TTR* calculation problem focuses on the second type of inaccuracy, because arriving peers push contact and resource information upon arrival. Our proposal, which we will call the *ATTR schema*, outperforms the adaptive method of reference, *AIMD*, by taking into account the latencies between the registered peer and the registration destination. This is achieved in two ways: low latencies are taken into account to adjust the *TTR* additive increase, and the new *TTR* is reduced when latencies exceed their average, due to expected peer losses, enabling earlier detection of peer outages.

The adaptive *TTR* (*ATTR*) values are calculated at the peers that receive the SIP REGISTER message (hereafter, *registrars*). The refresh procedure of each sender peer or resource (hereafter, *registry*) sets its associated timer accordingly. This approach allows refresh times to evolve independently. The registrar computes the *ATTR* values and inserts them in the "SIP 200 (OK)" message (received message is correct) in the *Contact* field (as the dSIP implementation does with other attribute-value pairs, e.g., peer-ID). All SIP messages modified by the *ATTR* schema are SIP-compliant. Peer and resource information is not replicated at the same peers in the overlay (e.g., peer information can be registered at peers *i*, *j* and *k*, and resource information can be registered at peers *x*, *w* and*z*).

The *ATTR* schema defines the following parameters:

- Initial refresh time ($T_{init}$). Initial refresh time value.
- Permanency time ($T_{perm}$). Time from the moment the registry joined the overlay. When a peer leaves the overlay, this parameter is reset to zero.
- Registration latency ($Lat_{reg}$). Time from the moment a registry sends a SIP REGISTER message to the point at which the registrar receives the request.
- Mean registration latency ($Lat_{mean\_reg}$). Average latency for each registry based on the registration latency computed by the registrars in the overlay for each registration message they receive.

Because the refresh time is calculated at the registrars (local refresh time), the confirmation response must contain the *ATTR* to be used by the registries. This way, information about the same peers or resources is registered in the overlay with different expiration times, depending on the latency between peers.

In a multi-hop network, delays between source and destination can be considerable and depend on the number of hops and the link delays. A sudden increase in delay may mean that a link will fail soon. The refresh time calculated in these conditions will therefore be short (the reduction will depend on the ratio between the current delay and the average delay).

Specific *ATTR* values are calculated from the $Lat_{mean\_reg}$ historical record and the time that the registrar obtains from the SIP REGISTER messages. For this purpose, the registry adds a timestamp to the original SIP message (the same procedure as for the *ATTR* value in the SIP 200 message), and the registrar obtains the latency by comparing the timestamp with its current time (the registrar is aware of the registry identifier or peer-ID). Thus, to achieve correct latency measurements, the overlay must be synchronized (either with a distributed network time protocol or with GPS, which is widely available in current mobile terminals). The registrar also records the registry (or resource) in a table with the expiration time $TTE = 2 \times ATTR$. Later, as previously mentioned, the registrar answers the SIP REGISTER message with an SIP 200 (OK) message (which we have also modified) that includes the *ATTR* value it calculated.

Peer and resource information is registered in the overlay according to a replication parameter ($r$), such that a registry sends $r$ SIP REGISTER messages to the overlay and receives $r$ SIP 200 (OK) messages from the registrars. The registry executes Algorithm 1 to update its *TTR* initial values with the maximum *ATTR* value received. This ensures that peer or resource information will not be in the overlay when the associated timer expires.

The schema computes *ATTR* values as:

$$\mathrm{ATTR}_j^i = \max\left(T_{init} + \frac{\log(T_{perm}^i)}{\log(1+1/T_{init})} - F_{lat,j}^i, T_{init}\right) \quad (3)$$

where 0< $i \le N_{peers}$ and 1< $j \le N_{resources}$ + 1. The main elements of the formula are:

- Maximization *versus* $T_{init}$, so $ATTR \ge T_{init}$. Thus, the overhead of the *ATTR* schema is upper-bounded by that of an original schema with (a fixed) $TTR = T_{init}$.

- The additive component log( Tiperm)/log(1 + 1/$T_{init}$) determines an *ATTR* logarithmic growth step when the nodes do not experience any trouble. By contrast, *AIMD* applies a constant growth step. Hence, registries that stay in the overlay for long will have lower *TTR* values under the *AIMD* schema, yielding excessive overhead.

- Filat,j represents the influence of delay on refresh time calculation. This subtractive component plays a role similar to that played by multiplicative decrements in *AIMD* under congestion. However, the *ATTR* schema is responsive to latency excess and not to congestion. In other words, ATTR values decrease only when the target peer is unusually hard to reach (an outage is expected) and not when it is subject to congestion.

The latency excess response parameter is computed as:

$$F_{lat,j}^i = \begin{cases} \exp\left(\dfrac{2Lat_{reg,j}^i}{Lat_{mean\_reg,j}^i}\right) & \text{if } Lat_{reg,j}^i > Lat_{mean\_reg,j}^i \\ F_{lat,j}^i \cdot tune & \text{otherwise, with } 0 < tune < 1 \end{cases} \quad (4)$$

where the first part is exponential to respond rapidly to above-average latencies and the second part is linearly responsive to below-average latencies. This second term is multiplied by a *tune* parameter to configure the speed of recovery from spurious delay increments. The *ATTR* schema follows the philosophy of Kademlia in the sense that it depends on permanency time, because the oldest peers are most likely to stay alive [16]. It is also parametric and depends on the initialization parameter, $T_{init}$, and the recovery parameter, *tune*. The excess latency response factor, $F_{lat}$, allows the schema to adapt to network conditions. As there is a different timer for each resource, as well as for each peer, there are fewer management message bursts than in the original schema. When a *TTR* timer expires, the associated peer and resource data are re-registered. The corresponding registration messages contain timestamps, and the registrar responses contain the *ATTR* values calculated with Equation (3). The registry will choose the largest of these values. This ensures that, given *r* registrars, the TTE timer of the best registrar (with the smallest latency) will not expire, even though the other timers may. This way, registration refresh is achieved at the minimal overhead cost.

When all the updating values received lie below threshold $T_{init}$, all other resource and peer information is re-registered with that minimum *ATTR* value, since a timer updating value below the threshold may indicate that a registry is likely to leave the overlay. This prevents stale information from remaining in the overlay for too long. Registries flush their registrations preventively when they find themselves likely to fail.

When remote registrars send low update values in their SIP 200 (OK) messages with low *ATTR* values, they are ignored by the registry, which instead chooses registrars with higher update values, because they are more likely to stay in the overlay (the probability of ungraceful departure is low). When all the *ATTR* values received are low, the probability of ungraceful departure is high. If the selected (highest) *ATTR* is below the threshold, the registry will automatically re-register itself (peer-ID) and all its resources with an associated *ATTR* that exceeds the threshold. This special registration uses SIP REGISTER messages with the defined minimum refresh time value, $T_{init}$.

**Algorithm 1:** Peer/resource registration update in the overlay.

```
r, replication parameter, α, concurrency parameter
n, number of SIP REGISTER messages sent
timer_j, the refresh timer for resource_j or peer information, is set to T_init initially
if timer_j expires then
  select resource_j or peer information
  set n = 0
  while n < r do
    if α > (r − n) then
      x = α
    else
      x = r − n
    end if
    select the x closest peers to resource-ID_j or peer-ID (not previously selected) from the DHT
    send x SIP REGISTER messages (resource_j or peer information)
    n = n + x
    while x > 0 do
      wait for SIP 200 (OK) or SIP REGISTER timeout
      decrement x
      if SIP 200 (OK) then
        timer_aux ← get ATTR from message(SIP 200 (OK))
        if timer_j < timer_aux then
          timer_j = timer_aux
        end if
      end if
    end while
  end while
  if timer_j ≤ T_init then
    timer_j = T_init
    Re-register the remaining resources and peer information
    (if timer_k > T_init) with a defined TTR value (T_init).
  end if
end if
```

# 5. ANALYSIS

To evaluate message overhead due to overlay maintenance in the original schema (periodic refresh times), we first assume that there is no churn. Let us consider that each peer is responsible for $N_{resources}$ resources and uses iterative lookup. According to [5], for a periodic refresh time, the average number of management messages per minute is:

$$N_{messages} \leq [2 \cdot (k \cdot (N_{resources} + 1) + (\log_b(N_{peers} - 1) + c) \cdot N_{resources})] \cdot ratio$$

where *ratio* is defined as the number of refresh periods per minute. Equivalently, *ratio* = 1/*TTR*. Therefore, both*AIMD* and *ATTR* need fewer management messages, because they allow increases in *TTR*. Each peer sends$N_{resources}$ +1 registration messages (representing itself and its resources) to the *r* closest peers. It also receives the same number of registration messages from the peers in the overlay, so it has to answer them. The iterative lookup to find suitable peers for $N_{resources}$ + 1 registrations requires up to $log_b(N_{peers} - 1) + c$ messages, where *c* is a small constant.

Figure 1 shows the average number of management messages per minute in the original schema for different population sizes (10, 100, 1,000, 10,000 and 100,000 peers in the overlay), with *r* = 3. The number of peers has less impact on the number of management messages than the replication factor. Signaling load is unacceptable for short refresh times.

However, if there is churn, for example, if the peers belong to an unstable wireless network, they will leave ungracefully with a certain probability, $p_{term}$ (defined for a one-minute time frame). Peer departure from the overlay is not notified, and the average time between departure and stale information expiration is a function of the remaining *TTR* and the expiration timer. The remaining *TTR* is:

$$\overline{TTR_{rem}} = \int_0^{TTR} t f_{term}(t)dt \tag{5}$$

where $f_{term}(t)$ is the distribution of ungraceful departures during the *TTR* cycle. Assuming a uniform distribution:

$$\overline{TTR_{rem}} = \frac{TTR}{2} \tag{6}$$

The expiration timer is originally set to *TTE* = 2 × *TTR*, so at the moment of departure, the time remaining to resource expiration is:

$$\overline{TTE_{rem}} = TTE - (TTR - \overline{TTR_{rem}}) = \frac{3}{2}TTR \tag{7}$$

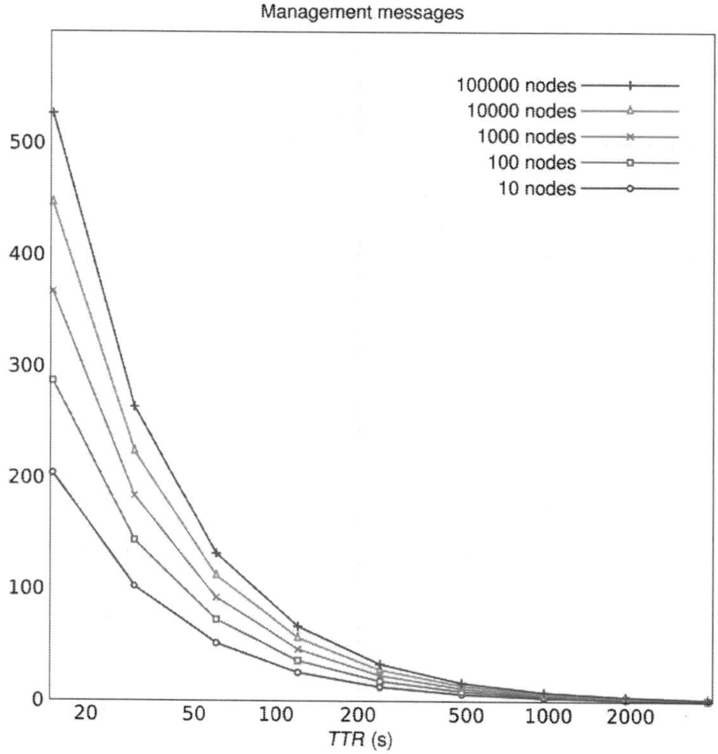


**Figure 1.** Number of management messages sent per minute for different refresh periods and population sizes, original schema.

Meanwhile, the stale information will remain registered in the overlay.

In addition to reducing management load, adaptive refresh times also decrease the probability of stale information due to ungraceful departures. The *ATTR* schema updates the refresh time with the highest value from the *r* SIP 200 (OK) messages received. If the registry is far from the registrars, the delay measured using SIP

REGISTER arrival times and timestamps will be long. $F_{lat}$ will also be high, and the refresh time will be short. In addition to reducing the probability of stale information, the time before stale registered information expires is parametric in *ATTR*, and it can be computed using Equation (7), assuming that the latency-based departure anticipation mechanism succeeds in making $ATTR = T_{init}$ iust before the event:

$$\overline{TTE_{proposal}} = \frac{3}{2} T_{init} \tag{8}$$

Therefore, thanks to a detection procedure for unannounced departures, stale information about peers that have left ungracefully only remains in the overlay for a short time, adjustable by the initial refresh time. As the peers do not refresh all the management data periodically, the probability of management information exchange between peers during ungraceful departure increases and the time that stale information remains in the overlay will tend towards the ideal limit defined in Equation (8).

In contrast to our proposal, the *AIMD* schema does not consider delays between peers (it only employs congestion estimation) and, therefore, does not anticipate ungraceful departures. Moreover, the *AIMD* schema continuously performs refresh procedures for all resources at the same time. This means that management messages travel in bursts. The *ATTR* schema only behaves in this way when an ungraceful departure is predicted from a latency excess. In addition, congestion is estimated less frequently in *AIMD* than latency in the *ATTR*schema, increasing the chance that in the *AIMD* system, departures will occur during higher values of $\overline{\text{TTE}}$ than in the *ATTR* schema. Therefore, with the *AIMD* schema, stale information remains in the overlay for longer than with the *ATTR* schema.

# 6. SIMULATION

## 6.1. Simulation Set-up and Merit Metrics

The OMNeT++ network simulator [52] was employed for the tests. The scenario we considered was a 100 × 100 m$^2$ square area with different peer population sizes, ranging from four to 100 randomly placed nodes. All the peers were mobile, with speeds of between eight and 20 m/s, following a Gaussian distribution. The underlying wireless *ad hoc* network was an IEEE 802.11s mesh. The simulations were repeated with 100 different random seeds for different population sizes. Each simulation lasted 3.600 s.

We simulated the original schema, *AIMD* [50] and our proposal to determine which had the lowest overhead of management messages for a comparable probability of stale information. We must emphasize that the logical links between overlay nodes were the same for the three schemas, since we did not notify the routing management in Kademlia. DHT information was always registered in the same nodes regardless of the schema we used. It is important to recall that, in the original schema, the probability of stale information in the overlay increases with *TTR*, as shown in Figure 2, according to:

$$p_{stale\_information} = 1 - (1 - p_{term})^{\overline{TTE_{rem}}/60} \quad (9)$$

where $p_{term}$ is the probability of node departures occurring in one minute, and thus, $p_{stale\_information}$ is the joint probability of this event within a $\overline{TTE_{rem}}/60\ \mathrm{min}$ window.

The management messages considered were SIP REGISTER and SIP 200 (OK) messages from the dSIP protocol. The DHTs and resource tables were fed by processing those messages. The lookup procedure also used dSIP register messages.

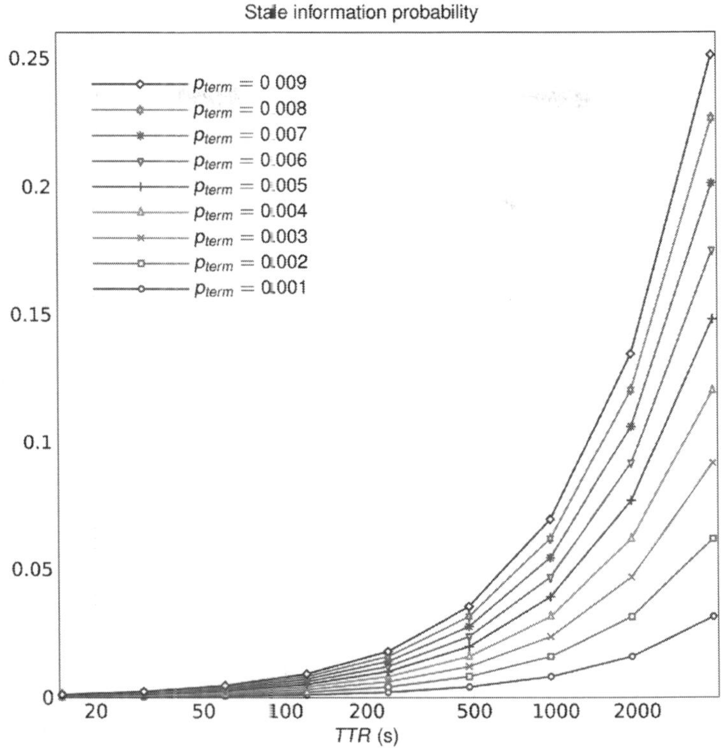


**Figure 2.** Probability of stale information in the overlay with the original schema.

- Moreover, the following assumptions were made:
- DHTs used iterative routing.
- Peers could leave or join the overlay (so there was churn).
- Peer identifiers were not uniformly distributed, and therefore, the number of entries in each *k*-bucket differed.

We adjusted the parameters of the original schema (*TTR*), *AIMD* ($T_{init}$, *C* and *D*) and *ATTR* ($T_{init}$, *tune*) for the same probability of stale information. This was done by inspection. The initial *TTR* was 15s for the original schema, so the peers performed four refresh actions per minute. $T_{init}$ was 15s for the *AIMD* and *ATTR* schemas, and thus, the mean permanency time of stale information with *AIMD* was the same as with our proposal. The replication and concurrency factors were three. Each peer registered three resources in the overlay. For *AIMD* *C*and *D*, we chose values that would allow moderate capacity variations, in order to achieve a probability of stale information comparable to that of *ATTR*. High *C* and *D* values would make it difficult for *AIMD* to determine that a node has left the overlay. Finally, the *tune* parameter was set to 0.875 for *ATTR* to keep the same probability of stale information as with the original schema. Availability was practically the same in the three scenarios. The examples in Section 6.2 correspond to $p_{stale\_information} = 10^{-3}$.

## 6.2. Comparison

Figure 3 compares the results of the schemas analyzed, for a confidence interval of 95% and a tolerance interval of 1%. The average number of management messages per minute with *ATTR* was much lower than with the original schema and *AIMD*, even for overlays with a few nodes. This difference became more significant as the number of nodes grew.

The *ATTR* schema achieved considerable improvements in terms of management overhead (approximately 21.50% of the number of maintenance messages in the original schema). Furthermore, *ATTR* required half the management messages of *AIMD*. We can illustrate this with an example. The average size of a management message in the Kademlia + dSIP implementation by the University of Parma is 689 bytes. Other P2PSIP implementations, such as RELOAD [53], have even larger average management message sizes. Using the original schema with a 100-node overlay, the management traffic rate is 26.182 kbps. With the *AIMD* schema, this rate is 16.995 kbps, and with our schema, it is only 10.565 kbps.

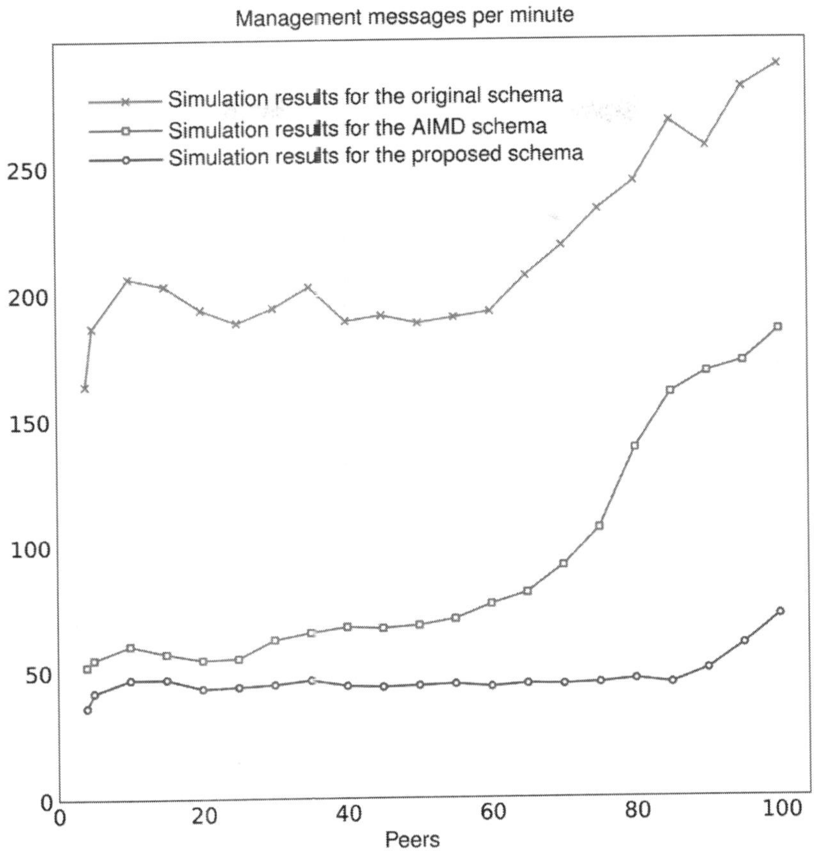


**Figure 3.** Comparison of management overhead for different population sizes (from 4 to 100 peers). Initial refresh time of 15 s for all schemas ($p = 0.001$.).

In addition to the results in Figure 3, the simulations also checked $p_{stale\_information}$. We verified that its value was ~$10^{-3}$ for the three schemas, so they were correctly tuned. Indeed, the probability of stale information was lower in all cases with *ATTR* than with *AIMD*. The information about the peers that left the overlay ungracefully persisted only for a short time (upper-bounded by the *time-to-refresh* of the original schema).

# 7. CONCLUSIONS

Wireless networks are evolving towards all-IP solutions that will rely on SIP for management purposes and sensor information exchange. In these networks, SIP technologies may follow a P2P approach, to enhance availability, storage and processing resources and scalability. In this context, P2PSIP protocols allow decentralized and fast communications. However, P2P mobile sensing overlays on top of wireless networks require further improvements in terms of management overhead and battery lifetime, which are closely related.

In this paper, we have presented a novel approach for reducing the traffic of management messages needed to maintain mobile sensing overlays on top of wireless networks, considering a Kademlia + dSIP P2PSIP implementation.

In the original Kademlia schema, based on periodic maintenance, short maintenance periods have a clear negative impact on overlay performance, since they increase management overhead. Furthermore, long maintenance periods increase the probability of stale information, which can lead to routing failures or longer lookup times due to timeouts.

We have proposed a new adaptive *TTR* (*ATTR*) schema that reduces management traffic while keeping the probability of stale information in the overlay at acceptable levels. Our simulations show that, for the same probability of stale information, our proposal outperforms both the original schema with fixed periodic refresh times and *AIMD*, one of the best dynamic *TTR* alternatives so far.

Our solution reduces management traffic and, as a consequence, it also mitigates link overload and bottlenecks. In forthcoming work, we will evaluate overload decrease in real mobile mapping and routing (MMR) networks, in which peers are aware of the topology. This configuration will allow the creation of logical links according to hop counts or delay measures and achieve faster lookups and, therefore, improved performance.

## ACKNOWLEDGMENTS

This research has been supported by projects Mefisto (10TIC006CT), Xunta de Galicia, Spain, and CALM (TEC2010-21405-C02-01), Mineco, Spain.

## REFERENCES

1. Chung, H.H.; Wu, Q. Managing Heterogeneous Wireless Sensor Networks with the Session Initiation Protocol (SIP). Proceedings of 14th International Conference on Advanced Communication Technology (ICACT 2012), Pyeong Chang, Korea, 19–22 February 2012; pp. 1042–1045.

2. Kanter, T.; Osterberg, P.; Walters, J.; Kardeby, V.; Forsstrom, S.; Pettersson, S. The Media Sense Framework. Proceedings of Fourth International Conference on Digital Telecommunications, Chamonix/Mont Blanc, France, 29 April–4 May 2012; pp. 144–147.

3. Ratnasamy, S.; Stoica, I.; Shenker, S. Routing Algorithms for DHTs: Some Open Questions. Proceedings of the International Workshop on Peer-To-Peer Systems (IPTPS 2002), Cambridge, MA, USA, 7–8 March 2002; pp. 45–52.

4. Bryan, D. dSIP: A P2P Approach to SIP Registration and Resource Location. Available online: http://tools.ietf.org/html/draft-bryan-p2psip-dsip-00.html (accessed on 5 November 2013).

5. Cirani, S.; Veltri, L. A Kademlia-Based DHT for Resource Lookup in P2PSIP. Avaiable online: http://tools.ietf.org/html/draft-cirani-p2psip-dsip-dhtkademlia-00 (accessed on 5 November 2013).

6. Wauthy, J.-F.; Schumacher, L. Implementation and Performance Evaluation of A P2PSIP Distributed Proxy/Registrar. Proceedings of the 2007 International Conference on Next Generation Mobile Applications, Services and Technologies (NGMAST 2007), Cardiff, UK, 12–14 September 2007; pp. 119–124.

7. Heikkinen, M.V.J.; Luukkainen, S. Technology Evolution of Mobile Peer-to-Peer Communications. Proceedings of the 4th Annual International Conference on Wireless Internet, Maui, HI, USA, 17–19 November 2008; pp. 1–9.

8. Ou, Z.; Harjula, E.; Kassinen, O.; Ylianttila, M. Feasibility Evaluation of A Communication-Oriented P2P System in Mobile Environments. Proceedings of the ACM Mobility Conference, Beijing, China, 20–25 September 2009; pp. 1–8.

9. Kassinen, O.; Harjula, E.; Ylianttila, M. Suitability of DHT-Based Peer-to-Peer Session Initiation Protocol for Wireless Distributed Services. Proceedings of the 12th International Symposium on Wireless Personal Multimedia Communications, Miyagi, Japan, 7–10 September 2009.

10. Lua, K.; Crowcroft, J.; Pias, M.; Sharma, R.; Lim, S. A survey and comparison of peer-to-peer overlay network schemes. *IEEE Commun. Surv. Tutor.* **2005**, *7*, 72–93.

11. Rowstron, A.; Druschel, P. Pastry: Scalable, Decentralized Object Location and Routing for Large-Scale Peer-to-Peer Systems. Proceedings of the IFIP/ACM International Conference on Distributed Systems Platforms, Heidelberg, Germany, 12–16 November 2001; pp. 329–350.

12. Zhao, B.Y.; Huang, L.; Stribling, J.; Rhea, S.C.; Joseph, A.D.; Kubiatowicz, J.D. Tapestry: A resilient global-scale overlay for service deployment. *IEEE J. Sel. Areas Commun.* **2004**, *22*, 41–53.

13. Stoica, I.; Morris, R.; Liben-Nowell, D.; Karger, D.R.; Kaashoek, M.F.; Dabek, F.; Balakrishnan, H. Chord: A scalable peer-to-peer lookup protocol for Internet applications. *IEEE/ACM Trans. Netw.* **2003**, *11*, 17–32.

14. University of Parma, Kademlia dSIP Implementation. Available online: http://www.mjsip.org/projects/p2psip/p2psip_dsip_071025.zip (accessed on 5 November 2013).

15. Meshkova, E.; Riihijarvi, J.; Petrova, M.; Mahonen, P. A survey on resource discovery mechanisms, peer-to-peer and service discovery frameworks. *Comput. Netw.* **2008**, *52*, 2097–2128.

16. Maymounkov, P.; Mazieres, D. Kademlia: A Peer-to-Peer Information System Based on the Xor Metric. Proceedings of the First International Workshop on Peer-to-Peer Systems (IPTPS 2002), Cambridge, MA, USA, 7–8 March 2002; pp. 53–65.

17. O'Connor, A.; Brady, C.; Byrne, P.; Olivre, A. *Characterising the eDonkey Peer-to-Peer File Sharing Network*; Trinity College: Dublin, Ireland, 2004.

18. Klimkin, A. *eDonkey Protocol Specification v0.6.2.* Available online: http://heanet.dl.sourceforge.net/sourceforge/pdonkey/eDonkey-protocol-0.6.2.html (accessed on 5 November 2013).

19. Liu, S.; Lei, W.; Pan, X.; Zhang, W. The Communication Model Based on SIP and ZigBee for Intelligent Home Electricity System. Proceedings of 4th International Conference on Multimedia Technology (ICMT 2011), Zurich, Switzerland, 27–28 June 2011; pp. 5594–5597.

20. Román-Portabales, A.; Pérez-Carrera, E.; González-Castaño, F.J.; Chaves-Diéguez, D. IMS Signaling for Smart Grid Home

Controllers. Proceedings of 19th IEEE International Conference on Consumer Electronics (ICCE 2011), Zurich, Switzerland, 11–15 September 2011; pp. 541–542.

21. Zhou, J.; Li, C.; Zhang, Z. Intelligent Transportation System Based on SIP/ZigBee Architecture. Proceedings of 2011 International Conference on Image Analysis and Signal Processing (IASP 2011), Wuhan, China, 21–23 October 2011; pp. 405–409.

22. Chen, M.; Yang, L.T.; Kwon, T.; Zhou, L.; Jo, M. Itinerary planning for energy-efficient agent communications in wireless sensor networks. *IEEE Trans. Veh. Technol.* **2011**, *60*, 3290–3299.

23. Chang, P.-H.; Wang, T.-P. Supporting Personal Mobility with Integrated RFID in VoIP Systems. Proceedings of International Conference on New Trends in Information and Service Science (NISS 2009), Beijing, China, 30 June–2 July 2009; pp. 1353–1359.

24. Rhee, J.; Park, H.; Bahg, Y.-J. Mobile Service Provision System Based on Sensor and SIP Event Notification Mechanism. Proceedings of the 2004 Joint Conference of the 10th Asia-Pacific Conference on Communications and the 5th International Symposium on Multi-Dimensional Mobile Communications, Beijing, Chian, 29 August–1 September 2004; pp. 740–743.

25. Lei, L.; Zhong, Z.; Lin, C.; Shen, X. Operator controlled device-to-device communications in LTE-advanced networks. *IEEE Wirel. Commun.* **2012**, *19*, 96–104.

26. Chiba, T.; Warabino, T.; Yokota, H. Implementation and Evaluation of Handoff Method between Access Nodes for P2PSIP and IMS Cooperating Networks. Proceedings of 19th International Conference on Computer Communications and Networks (ICCCN 2010), Zurich, Switzerland, 2–5 August 2010; pp. 1–6.

27. Kansal, A.; Goraczko, M.; Zhao, F. Building A Sensor Network of Mobile Phones. Proceedings of 6th International Symposium on

Information Processing in Sensor Networks (IPSN 2007), Cambridge, MA, USA, 25–27 April 2007; pp. 547–548.

28. Shen, C.; Shen, F.; Wu, Z.; Luo, J. Application of session initiation protocol to networked sensor interfaces.*Comput. Stand. Interfaces* **2009**, *31*, 454–457.

29. Xu, F.; Jin, H.; Liao, X.; Qiu, F. Enhancing the Reliability of SIP Service in Large-Scale P2P-SIP Networks. Proceedings of the 6th International Conference on Advances in Grid and Pervasive Computing (GPC 2011), Oulu, Finland, 11–13 May 2011; pp. 52–61.

30. Wang, A.; Heinzelman, W.R.; Chandrakasan, A.P. Energy-scalable protocols for battery-operated microsensor networks. *J. VLSI Signal Process. Syst. Signal Image Video Technol.* **2001**, *29*, 223–237.

31. Yick, J.; Mukherjee, B.; Ghosal, D. Wireless sensor network survey. *Comput. Netw.* **2008**, *52*, 2292–2330.

32. Lv, Q.; Cao, P.; Cohen, E.; Li, K.; Shenker, S. Search and Replication in Unstructured Peer-to-Peer Networks. Proceedings of the 16th annual ACM International Conference on Supercomputing, New York, NY, USA, 22–26 June 2002; pp. 84–95.

33. Fletcher, G.H.L.; Sheth, H.A.; Brner, K. Unstructured Peer-to-Peer Networks: Topological Properties and Search Performance. Proceedings of Third International Joint Conference on Autonomous Agents and Multi-Agent Systems. W6: Agents and Peer-to-Peer Computing, New York, NY, USA, 19 July 2004; pp. 14–27.

34. Morselli, R.; Bhattacharjee, B.; Srinivasan, A.; Marsh, M.A. Efficient Lookup on Unstructured Topologies. Proceedings of the 24th Annual ACM Symposium on Principles of Distributed Computing, Las Vegas, NV, USA, 17–20 July 2005; pp. 77–86.

35. Zhang, Y.; Liu, W.; Lou, W.; Fang, Y. Location-based compromise tolerant security mechanisms in wireless sensor networks. *IEEE J. Sel. Areas Commun.* **2006**, *24*, 247–260.

36. Stutzbach, D.; Rejaie, R. Improving Lookup Performance over A Widely-Deployed DHT. Proceedings of the 25th IEEE International Conference on Computer Communications, Barcelona, Spain, 23–29 April 2006; pp. 1–12.

37. Rosenberg, J.; Schulzrinne, H.; Camarillo, G.; Johnston, A.; Peterson, J.; Sparks, R.; Handley, M.; Schooler, E. SIP: Session Initiation Protocol. Available online: http://www.ietf. org/rfc/rfc3261.txt (accessed on 5 November 2013).

38. Li, J.; Stribling, J.; Gil, T.M.; Morris, R.; Kaashoek, M.F. Comparing the Performance of Distributed Hash Tables under Churn. Proceedings of the Third International Workshop on Peer-To-Peer Systems (IPTPS 2004), La Jolla, CA, USA, 26–27 February 2004; pp. 87–94.

39. Androutsellis-Theotokis, S.; Spinellis, D. A survey of peer-to-peer content distribution technologies. *ACM Comput. Surv.* **2004**, *36*, 335–371.

40. Oechsner, S.; Hossfeld, T.; Tutschku, K.; Andersen, F.U.; Caviglione, L. Using Kademlia for the Configuration of B3G Radio Access Nodes. Proceedings of the Fourth Annual IEEE International Conference on Pervasive Computing and Communications Workshops, Lugano, Switzerland, 19–23 March 2012; pp. 141–145.

41. Cholez, T.; Chrisment, I.; Festor, O. A Distributed and Adaptive Revocation Mechanism for P2P Networks. Proceedings of the Seventh International Conference on Networking, Cancun, Mexico, 13–18 April 2008; pp. 290–295.

42. Kassinen, O.; Harjula, E.; Korhonen, J.; Ylianttila, M. Battery Life of Mobile Peers with UMTS and WLAN in A Kademlia-Based P2P

Overlay. Proceedings of the 20th Personal, Indoor and Mobile Radio Communications Symposium, Tokyo, Japan, 13–16 September 2009; pp. 689–705.

43. Zhao, B.; Wen, Y.; Zhao, H. KDSR: An Efficient DHT-Based Routing Protocol for Mobile *Ad Hoc* Networks. Proceedings of the Ninth International Conference on Hybrid Intelligent Systems, Shenyang, China, 12–14 August 2009; pp. 245–249.

44. Sotiriou, A.-D.; Kalliaras, P.; Mitrou, N. Flexible routing in a distributed K-ary tree: The K-Umbrella. *Eur. Trans. Telecommun.* **2009**, *20*, 139–157.

45. Rhea, S.; Geels, D.; Roscoe, T.; Kubiatowicz, J. Handling Churn in A DHT. Proceedings of 2003 USENIX Annual Technical Conference, San Antonio, TX, USA, 9–14 June 2003; pp. 10–11.

46. Koskela, T.; Kassinen, O.; Korhonen, J.; Ou, Z.; Ylianttila, M. Peer-to-Peer Community Management Using Structured Overlay Networks. Proceedings of 2008 International Conference on Mobile Technology, Applications and Systems, Innsbruck, Austria, 10–12 September 2008; pp. 1–6.

47. Maenpaa, J.; Camarillo, G. Study on Maintenance Operations in A Chord-Based Peer-to-Peer Session Initiation Protocol Overlay Network. Proceedings of 2009 IEEE International Symposium on Parallel and Distributed Processing, Rome, Italy, 23–29 May 2009; pp. 1–9.

48. Chan, H.N.; Van, K.N.; Hoang, G.N. Characterizing Chord, Kelips and Tapestry Algorithms in P2P Streaming Applications Over Wireless Network. Proceedings of the Second International Conference on Communications and Electronics, Hoi An, Vietnam, 4–6 June 2008; pp. 126–131.

49. Kelenyi, L.; Nurminen, J.K. Optimizing Energy Consumption of Mobile Nodes in Heterogeneous Kademlia Based Distributed Hash Tables. Proceedings of the 2nd International Conference

and Exhibition on Next Generation Mobile Applications, Services and Technologies, Cardiff, UK, 16–19 September 2008; pp. 70–75.

50. Liu, X.; Lan, J.; Shenoy, P.; Ramaritham, K. Consistency maintenance in dynamic peer-to-peer overlay networks. *Comput. Netw.* **2006**, *50*, 859–876.

51. Martin, S.; Leduc, G. A Dynamic Neighborhood Discovery Protocol for Active Overlay Networks. Proceedings of the IFIP TC6 5th International Workshop in Active Networks, Kyoto, Japan, 10–12 December 2003; pp. 151–162.

52. OMNeT++ Simulator. Available online: http://www. omnetpp.org (accessed on 5 November 2013).

53. Jennings, C.; Lowekamp, B.; Rescorla, E.; Baset, S.; Schulzrinne, H. Resource Location and Discovery. Available online: http://www.p2psip.org/drafts/draft-ietf-p2psip-reload-00.txt (accessed on 5 November 2013).

# INDEX

3GPP, 92, 94, 105, 113, 114, 232, 233, 236, 241, 242, 243, 244

## A

Artificial speech quality assessors, 32
ATTR Schema, 263

## B

Bursty packet losses, 32

## C

CINR Level, 8
Cisco packet tracer, 116
Cluster, 56, 71, 72, 73, 76, 77, 78, 86

## E

E-Model, 29, 39, 55, 56, 57, 58, 62, 68, 70
E-UTRA Overview, 199

## F

Frame Retries, 6, 28

## G

Genome, 43, 44, 46, 55, 56, 57, 60, 62

## H

Handover Management, 1, 2, 28, 29
Handover Triggers, 2, 5

## K

Kademlia, 250, 252, 253, 256, 257, 258, 259, 260, 261, 263, 266, 272, 274, 276, 278, 279, 282, 283

## L

Landmark node, 77

## M

Mobile communication systems, 196

## O

One Voice, 229, 230, 231, 238, 241, 242, 243, 244, 245, 246, 247
Ordinary node, 77

## P

P2P overlays, 250, 252, 255
P2P Overlays, 255
PDCCH Overview, 202
Peer-to-Peer VoIP, 71, 72
Proactive controller, 143, 147
Prototyping, 116, 120

## Q

Q-Model, 44, 45, 54, 56, 57, 58, 62
QoE, 32

## R

Rationale, 148, 149, 151, 154
Relay Lookup, 71, 72, 76
Relay node, 77

## S

Single/Bi-Casting, 2, 16
SRVCC, 230, 239, 244, 247

## T

Telephony, 67, 115, 116, 118, 122, 130, 247
Time domain priority, 155
Traffic Model, 103, 211

## V

Voice over LTE, 230, 233, 238, 244, 246, 247
VoIP, 1, 2, 3, 4, 5, 6, 7, 8, 9, 10, 227, 233, 249, 250, 252, 254, 280
VoIPv6, 179, 180, 181, 183, 191, 192, 193, 194